Congestion

POLITICAL AND SOCIAL CHANGE

Edited by Martin Bak Jørgensen and Óscar García Agustín

VOLUME 2

Zu Qualitätssicherung und Peer Review der vorliegenden Publikation

Die Qualität der in dieser Reihe erscheinenden Arbeiten wird vor der Publikation durch einen Herausgeber der Reihe sowie durch einen externen, von der Herausgeberschaft ernannten Gutachter im Blind-Verfahren geprüft. Dabei ist der Autor der Arbeit dem Gutachter während der Prüfung namentlich nicht bekannt.

Notes on the quality assurance and peer review of this publication

Prior to publication, the quality of the work published in this series is reviewed by one of the editors of the series and blind reviewed by an external referee appointed by the editorship. The referee is not aware of the author's name when performing the review.

Laura Bang Lindegaard

Congestion

Rationalising Automobility in the Face of Climate Change

PETER LANG
EDITION

Bibliographic Information published by the Deutsche Nationalbibliothek
The Deutsche Nationalbibliothek lists this publication in the Deutsche Nationalbibliografie; detailed bibliographic data is available in the internet at http://dnb.d-nb.de.

Library of Congress Cataloging-in-Publication Data
Lindegaard, Laura Bang, 1980- author.
Congestion : rationalising automobility in the face of climate change / Laura Bang Lindegaard.
pages cm — (Political and social change ; volume 2)
Includes bibliographical references.
ISBN 978-3-631-66663-0
1. Transportation, Automotive—Social aspects. 2. Transportation—Social aspects. 3. Transportation—Environmental aspects. 4. Climatic changes—Social aspects. I. Title.
HE5611.L56 2015
303.48'32—dc23

2015027571

Cover illustration:
Cars stink © Peter Blanchard
http://www.peterblanchard.ca/

ISSN 2198-8595
ISBN 978-3-631-66663-0 (Print)
E-ISBN 978-3-653-06098-0 (E-Book)
DOI 10.3726/978-3-653-06098-0

Peter Lang Edition is an Imprint of Peter Lang GmbH.

Peter Lang – Frankfurt am Main · Bern · Bruxelles · New York · Oxford · Warszawa · Wien

This publication has been peer reviewed.

www.peterlang.com

Acknowledgements

First of all, I would like to thank the citizens of Horslunde who participated in my empirical study. Without you this book would not have been. Not only is the book dependent on the fact that you spent afternoons and evenings patiently participating in my focus groups; it is also dependent on your enthusiasm and warmth in doing so. Furthermore, I am grateful to my colleagues at Aalborg University, particularly my supervisors Paul McIlvenny and Inger Lassen, and to my husband, Henrik. Without all of you, I would not have been able to complete the work with this book.

Table of Contents

1. Introduction

When the Model T was launched, the owner of Ford Motors, Henry Ford, is known to have remarked that the consumers could get it in any colour they wanted as long as they wanted it in black. Ford's Model T was the first automobile produced on moving assembly lines with completely interchangeable parts, and the introduction of this new mode of production is widely considered to have democratised automobility. With the reduced production costs, automobility was no longer restricted to the economic elite but affordable to the growing American middle class. Hence, even though the Model T was a standardised product that left the consumer with few alternatives, it quickly came to epitomise individual autonomy. Thanks to the wonder of assembly line production, the American consumer could get what he (or, rarely, she) wanted and go where he wanted, unrestricted by route plans and time schedules. He could go *On The Road* as Jack Kerouac famously did in the 1950s, and as generations have done ever since. Indeed, the fantasy of unhindered automobility extends well into the future where the automobile has transcended the last infrastructural restrictions on individual mobility and colonised the sky. Consider, for instance, the stark contrast between the streets below and the sky above the city of Los Angeles in Ridley Scott's science fiction classic *Blade Runner*. While the streets of Los Angeles are crowded by an impassable *mélange* of people, objects and cultures, flying automobiles move unimpededly in the uncrowded sky above.

Today, a century after the introduction of the Model T, automobility has proved to be a severe, if not unsurpassable problem for contemporary Western societies. The physical infrastructure in larger cities is under constant pressure, and expansions of the existing capacity simply generate additional traffic and relocate the problem to a not so distant future. Ironically, then, the fascination with 'the flying car' in twentieth-century science fiction is not simply an extrapolation of the automobile's technological development, but also, and perhaps foremost, an extrapolation of an inherent antagonism within automobility and a desire for a neat technological fix. Indeed, if automobility were in fact universalised, then the pursuit of individual mobility would probably terminate in infrastructural congestion and collective immobility. In addition to what are, after all, merely practical problems, automobility is intimately connected with contemporary societies' dependency on fossil fuels and, by implication, the current climate crisis. According to Van der Hoeven et al. (2012), the transport sector accounted for twenty-two per cent of the world's energy-related CO2 emissions in 2010, and road transportation accounted for almost three quarters of the transport sector's overall emissions, and it is generally acknowledged that automotive mobility is a harmful co-constituent in the current crises (for instance, Böhm et al., 2006; Paterson, 2007; Urry, 2008 and 2011). Yet, automobility is still a, if not the, preferred mode of transportation in the West, and there is little to indicate that this will change in the foreseeable future. According to Urry (2004), Western societies are "locked in" in "systems of automobility", that is, "self-organizing, self-generating, non-linear world-wide systems of cars, car-drivers, roads, petroleum supplies, and many novel objects, technologies, and signs" (27), and change will only happen if a whole range of minor, unpredictable turning points happen to happen at the same time, thereby causing a tipping point that will lead to new systems after the car (27).

Acknowledging the relevance of material configurations, this book suggests that social change (or the lack thereof) is not simply a consequence of the blind and unpredictable happenstance of complex systems, but also of traceable governmental rationalities at all levels. Drawing on both studies of governmentality and ethnomethodology, the book adds an as yet underdeveloped perspective to already existing research on automobility: it adds to the relatively few qualitative enquiries into the interdependency of current forms of mobility and climate change by suggesting that in order to properly understand this interdependency, it is necessary to go beyond a systems-oriented approach to automobility and focus also on how the governing of car-dependent living is mutually constituted as the government of others and the government of the self intersect in discursive interaction. Accordingly, the book reports on an empirical study of the execution of a municipal transportation strategy in a small village in Denmark, aiming at 'greening' the everyday transportation practices of its citizens. Through meticulous analyses of focus group data stemming from sessions with villagers both before and after the execution of the municipal strategy, the book aims at demonstrating two interrelated phenomena. Firstly, the analysis of data stemming from before the transportation strategy focuses on demonstrating the subtle ways in which the villagers rationalise their transportation conduct. Secondly, the analysis of data stemming from after the execution of the strategy focuses at demonstrating the discursive accomplishment of the contested intersection of, on the one hand, the municipal aim to conduct the transportation conduct of the villagers, thereby turning them into 'green drivers', and, on the other hand, the villagers' resistance to having their conduct conducted. That is, the second part of the analysis shows that whereas the Municipality attempts to govern the citizens not by crushing their capacity to act, but by utilising it in getting the citizens to act differently (cf. Rose 1999), the citizens skilfully utilise this way of governing to rationalise their everyday practices, particularly, their transportation practices, as already being as sustainable as possible. Significantly, the analysis demonstrates how the villagers negotiate the municipal strategy in ways that merely sustain their memberships of 'sustainable movers' and, by implication, their car-dependent life-forms, and as such the book illuminates the ways in which our car-dependent living is bound up with a particular way of governing which relies on continuous processes of rationalisation.

Considering the blatant fact that automobility accounts for an extremely high amount of human-caused environmental pollution and CO2 emissions, it may be no surprise that energy use reduction and climate change mitigation currently feature prominently on the academic transport research agenda. Recent studies within this domain emphasise the need to broaden the research perspectives beyond narrow disciplinary boundaries in order to meet the challenges of the increasing climate and energy crises. For instance, both Schwanen et al. (2011) and Banister et al. (2012) speak from the point of view of transport studies, such as transport geography, but they both point out that transport research would benefit tremendously from insights, concepts and methods from other disciplines, the social sciences in particular. Considering such developments within transportation research, it may, on the contrary, be a slight surprise that there are still only relatively few studies focusing on energy use reduction and climate change mitigation within the more designated field of mobilities research and within the social sciences as such. However, there has been done some work concerning sustainable mobility (Dudley et al. 2011; Freudendal-Pedersen 2009), climate change and future fuel systems (Urry 2008 and 2011) and a little more concerning somehow critical automobility studies (Böhm et. al. 2006; Conley & McLaren 2009; Cresswell 2010; Paterson 2007; Scott 2013; Sheller & Urry 2000; Urry 2004).

Only a few of these works approach car-dependent living not prominently as an unpredictable, 'innocent' system, but rather as a *regime* that is mutually constituted by a particular way of governing dispersed at all levels. However, Böhm et al. (2009) provide a lucid argument that both acknowledges Urry's paradigmatic work on systems of automobility and, importantly, complements it. In more detail, the volume starts with the observation that whereas the growing literature on automobility has provided a heightened understanding of the phenomenon, the inherent political dimensions of the phenomenon have mostly been neglected. Accordingly, the ambition of the volume is not only to theorise automobility, but, critically, to politicise it through a move from systems to regimes of automobility. Firstly, Böhm et al. consider the term 'automobility', suggesting that the word invokes an understanding of "a patterned system which is predicated in the most fundamental sense on a combination of notions of autonomy and mobility" (4). Drawing on Laclau and Mouffe's concept of 'chains of equivalence' (Laclau and Mouffe, 1985), they point out that the relation between these aspects has been automaticised, even though it has no a priori necessity, and they stress that when they talk of a system of automobility, they talk of the somehow systematic and patterned way in which an array of social developments have contingently reinforced each other until a point when automobility seems not only possible but, in many instances, necessary (5). Secondly, however, Böhm et al. observe that the notion of a 'system' may provide an insufficient understanding of the contingent but powerful patterning of automobility. They point out that Urry's notion of a viral, self-organising, autopoietic and non-linear system (for instance, Urry, 2004: 27) neglects or leaves out important and inevitable aspects of the phenomenon of automobility, dramatically underestimating the importance of different forms of collective human agency and political power (5). On the contrary, Böhm et al. suggest that it may be more fruitful to speak of *regimes* of automobility in that the notion of regimes makes it possible to maintain an orientation to automobility as somehow systematic (that is, as it will continuously be unfolded, capable of rationalising itself), but at the same time acknowledge and make available for enquiry the relations of power that co-construct the very sense of a patterned and necessary system (6). Hence, taking inspiration from Foucault, Böhm et al. unfold the notion of such a regime, proposing that it is co-constituted by the interrelated and mutually supportive elements of truth, power and subjectivity (6–9).

Also Paterson (2007) is concerned with automobility as being prominently a problem of governing through the conduct of conduct which is manifestly unsustainable in its current form (27). He stresses that contemporary automobility depends on a certain governmental power through which governments at one and the same time have succeeded in promoting and sustaining car-dependent transportation and in (re)producing the modern subject as 'autonomously mobile'. He argues that the construction of a decidedly automobile subject is the pre-condition of automobility, as the subject is the automobile's "principal daily agent" (165), and he emphasises that this produced subject takes on multiple forms and that its ongoing reproduction depends on the complex interplay of "popular cultural forms, daily practice, regulatory interventions, surveillance and resistance" (165). In this way, Paterson points to how the 'liberated' automobile subject is in fact produced and thus subjected to this liberation, thereby emphasising the double sense of automobility as simultaneously liberating and dominating (142). Accordingly, Paterson, too, is concerned with the lack of policy within Urryian systems theory. He observes that Urry's "nicely turned phrases that automobility is 'immensely flexible and wholly coercive', or that automobility 'coerces people into an intense flexibility'" (125) fail to pay attention to "the concrete decisions and the struggles over them which favoured automobility over its alternatives" (26) and beg the question of how it ended up this way (125). Hence, Paterson's overall point is that if we really want to understand how automobility came to be as it is and, ultimately, how we end it, it is necessary to move beyond a systems approach and acknowledge that automobility is inevitably bound up with governing the subject (223–49).[1]

Whereas I follow scholars such as Böhm et al. and Paterson and approach car-dependent living as prominently maintained through a particular way of governing the conduct of others and the self, I add an as yet underdeveloped perspective to such an approach, namely a perspective that focuses on how governmental rationalities of automobility are continuously negotiated in everyday discursive interaction. Specifically, the book proposes a perspective in which governmental rationalities are seen as co-constituted in the discursive negotiation and resistance of different politically infused strategies and projects, such as, in this case, a municipal transportation strategy aimed at conducting villagers' transportation conduct in a small village in Denmark.

By approaching automobility as a *problem* of governance, this book is imbued with insights from studies of governmentality. I understand the governance of automobility as an ongoing activity, dispersed into all levels of human interaction, and, by extension, I understand power in a Foucauldian way: not as an unambiguously repressive force, but as a productive force, continuously and creatively producing and reproducing its own conditions of possibility. In current studies of governmentality, Rose (1999) has taken inspiration from Foucault, and his work on the powers of freedom represents the backbone of my approach to automobility as a problem of governance. Focusing on *how* recent forms of governing are connected to a particular way of utilising freedom, itself a set of powers, Rose founds the understanding that runs through the book, namely that current forms of governing are co-constituted by certain rationalities of power

1 To my knowledge, Paterson is the only scholar approaching automobility as a problem of governance who is explicitly concerned with the question of greening or ending the car. Whereas Böhm et al. (2009) acknowledge such questions, broadly discussing the antagonisms of the car, none of the chapters really engage in how to 'green' or end the car. Also, whereas both Seiler (2008) and Packer (2008) discuss automobility in terms of Foucauldian notions of government and power, none of them are concerned with the questions of how automobility is bound up with climate change or sustainability.

that are continuously negotiated at all levels, importantly also in discursive interaction at the most vulgar, everyday level.

However, whereas scholars such as Rose indicate that governing is inextricably co-constituted in mundane, discursive interaction, none of them have carried out studies that make the *hows* of such interaction available for enquiry, and neither has anyone in the narrower field that focuses on automobility as a problem of governance. Consequently, I have found it fruitful to connect my governmentality perspective with ethnomethodology to be able to make enquiries into the yet underexamined governing that unfolds in the discursive interaction of ordinary people. Not only does ethnomethodology understand rationality and moral order in a way that can be said to be compatible with and clarifying the understanding within studies of governmentality, it also provides a unique starting point for analyses of the *hows* of this discursive interaction. Garfinkel (1967) sets the agenda for a line of research that focuses on the accomplishment of observably relevant tasks in specific situations, transposing what is otherwise thought of as philosophical and sociological problems into practical ones. Respecifying the classic problems of morality, Jayyusi (1984; 1991) not only expounds that rationality and morality are closely related phenomena accomplished in discursive interaction, she also indicates how this accomplishment can be made available for enquiry.

Hence, drawing on both studies of governmentality and ethnomethodology, the objective of the book is to contribute to the research of automobility as a problem of governance, adding a perspective to the already existing research, namely a perspective that enquires into *how* the rationalities of government are accomplished in discursive interaction at the vulgar, everyday level. Whereas studies of automobility as a problem of governance have so far attempted to focus on broader, genealogical investigations of the rationalities of government, this book reports on an in-depth qualitative study. The qualitative study focuses on the discursive accomplishment of the contested intersection of a municipal transportation strategy aimed at 'greening' citizens' everyday transportation practices and the self-administration of these citizens, and I demonstrate how the intersection is accomplished in situ in the citizens' accounts for their transportation practices. However, I do not want to invoke a persistent but problematic micro/macro dichotomy via this relative positioning against the already existing research. Rather, I draw on Latour (2002) and talk about the molecular and the molar as merely stretchable areas, intertwined in differently dispersed networks. Lastly, since this expected contribution to automobility research is premised on my attempt to draw on insights from ethnomethodology to make a first stage analysis of orders of rationalities that is then, secondly, interpreted in terms of governmentality theory, the book, by extension, also aims to contribute to the theoretical and methodological development of studies of governmentality.

Whereas the possible contributions to automobility research and to governmentality studies stem from my overall research question of *how* certain governmental rationalities are accomplished in talk-in-interaction, the book also aims at an arguably more normative contribution to the ending of the unsustainable regime of automobility. This is not an uncontroversial ambition for a book based on studies of governmentality and ethnomethodology, and my claim should already now be considered in slightly more detail. Richards (2005) mentions two models of research within ethnomethodological conversation analysis. The first model starts with the discovery of a social phenomenon "and on the basis of this prescribes – or proscribes – certain causes of actions" (3–4). The second model represents "the primary research as oriented predominantly to description" (5) and recognises this description's "potential for enriching professional practice rather than reframing it in prescriptive terms" (5). Discussing the models, Richards states that the first is antithetical to ethnomethodologically informed research due to its starting point in the discovery of a *particular* problem and due to the assumption that proscription and prescription imply "that it is possible to specify exactly what actors should do in particular circumstances" (4). However, I would like to question this position. Firstly, I find it difficult to see how any researcher could escape the 'moment' of discovery in which the researcher decides to do one kind of enquiry rather than another into one site of research rather than another, and, following Jayyusi (1991), I would suggest that this moment is necessarily *normative* insofar as it is theoretically and practically impossible to exit the moral order (247). Secondly, I find it difficult to see that such a moment of discovery needs to exclude the possibility of an analysis conducted in the mode of indifferent description. Hence, throughout the book I argue for a position that recognises that it is, after all, not a theoretical and methodological impossibility to acknowledge an inevitably normative position and to carry out a study that does not merely presuppose its own findings. In Foucauldian wording, it is not impossible to carry out a study with the ambition of asking: if this were really 'true', what would you do? (Foucault, 1995; cf. Biebricher, 2008).

The first part of the book comprises three chapters which together form the theoretical starting point. Whereas Chapters Two and Three introduce and discuss relevant foundations of and developments within studies of governmentality and ethnomethodology, respectively, the shorter Chapter Four concerns a more systematic account for the theoretical soundness of connecting the two. The chapter on governmentality focuses on the notion of freedom as a set of powers, on the related notion of the conduct of conduct and on the general research orientation to *how* questions. Furthermore, the chapter discusses the understanding that rationalities of government are continuously accomplished in discursive interaction, and it pursues the debate within current studies of governmentality regarding the importance of acknowledging that *resistance* is a constituent moment of governing. The chapter on ethnomethodology focuses on the praxiological understanding of rationality and morality as continuously negotiated, and accomplished, in talk-in-interaction. In the foundational discussions of rationality and morality, I further present the specific line of ethnomethodological research that I draw on in the book; a line that acknowledges the relation between Garfinkel's founding thinking in ethnomethodology and Sacks's original work in conversation analysis and membership categorisation analysis. Moreover, since theory and method are interconnected phenomena in ethnomethodology, it is almost impossible to keep the exposition nice and tidy, and the chapter is inherently intertwined with and prefigures methodological issues.

The second part of the book also comprises three chapters, concerned with methodology and analysis, respectively. The first chapter of the second part clarifies how I approach the discursive accomplishment of rationalities that co-constitute the governing of automobility, or, more specifically, how I approach the rationalisation and negotiation of the intersection of a municipal strategy aimed at 'improving' citizens' transportation practices and citizens' self-administration. Whereas the methods presented are predominantly informed by ethnomethodology, they are chosen and discussed with the particular purpose of enabling an analysis of orders of rationalities, interpreted via insights from studies of governmentality. The following two chapters comprise the analysis proper. Appropriately, the analysis is based on an ethnomethodological approach to how rational, moral ordering can be made available for enquiry in talk-in-interaction and on Dean's (2010) observation that the simultaneously rational and moral character of government is observable in people's *accounts* for their conduct. Both chapters provide in-depth analyses that demonstrate the *hows* of the accomplishment of rationalities that co-constitute the governing of automobility in the face of climate change. However, they differ, as the first focuses on how citizens' own transportation conduct is made rationally accountable, and the second focuses more specifically on the accomplishment of the contested intersection of a municipal strategy aimed at conducting citizens' transportation conduct and citizens' attempt to conduct their own transportation conduct.

2. Studies of governmentality

The present chapter comprises four sections. The first one concerns selected areas of Foucault's work, focusing on Foucault's work on the notion of governmentality. The remaining sections concern current studies of governmentality and depict a line of research which is of particular relevance in this context. The first, and lengthiest of these sections, concerns Rose's work on the powers of freedom. This is no coincidence but should reflect the fact that Rose's work represents the backbone of my approach to governmentality. By focusing on how recent forms of governing are connected to a particular way of utilising freedom as being itself a set of powers, Rose founds the understanding that runs through this book, namely that current forms of governing are co-constituted by certain mentalities or rationalities of power that are continuously negotiated in discursive interaction. However, whereas Rose represents the backdrop of the approach, I have found it useful to broaden the scope and draw on the work of other scholars to clarify some of Rose's points which are of specific relevance to my project. Hence, in the third section I draw on Dean in the aim to connect more clearly the notions of morality, mentalities of government and regimes of practices. Specifically, Dean explicates what Rose touches upon, namely that the specific notion of government implies that rationality is closely related to morality because government becomes a matter of rational accountability of practices concerned with 'conducting the conduct' of others and practices concerned with conducting one's own conduct. In the last sections, I turn towards some more critical issues within studies of governmentality, pointing to the most recent discussions that are of relevance to my book. Hence, in the fourth section, I revisit a possible inadequacy in the Foucauldian starting point, focusing on possible pitfalls for current studies and suggesting that these pitfalls can be avoided if studies of governmentality recognise that resistance must be dealt with as a constitutive moment of government. Therefore, the general purpose of the last section is to point out that not only is current governing co-constituted by certain 'regimes of rationality', that is, by certain ways of accomplishing meaningful accounts for the conduct of conduct, governing is a continuous and contested process in which rationalities are not simply maintained, but negotiated and resisted. Hence, when I talk about automobility as a problem of governance, it should invoke the recognition that governance is a continuously problematised activity, not only at a more 'overall' level of legislation, but at all levels, including the 'molecular' level where programmes for the administration of others intersect with techniques for the administration of ourselves.

The Foucauldian heritage

Foucault first introduces the concept of governmentality in his Collèges de France lectures in 1978 and 1979. He approaches the concept through a historical investigation of the emergence of liberal government in the eighteenth century. Foucault's history of governmentality is a study of the process through which the state of justice is governmentalised, and, importantly, through which the first descriptions of the art of government become possible. These studies grow from Foucault's reading of parts of the reception, critique and rebuttal of Machiavelli's *The Prince*. According to Foucault, this text generated all further discussions of government from the sixteenth century to the eighteenth century. Specifically, Foucault (2007) observes that whether or not the many anti-Machiavellian treatises represent an accurate understanding of Machiavelli's project, they do represent a radical new orientation to questions of government. Whereas Machiavelli's *The Prince* is considered to be about the prince's ability to keep his principality, the critiquing texts are about "something else and new, namely the art of government" (131).

Foucault focuses on one of the earliest texts of this "great anti-Machiavellian" (131) literature, namely Guillaume de La Perrière's *Miroir Politique*. In the present context it suffices to mention Foucault's discussion of La Perrière's statement "'government is the right disposition of things, arranged so as to lead to a convenient end'" (134). Especially Foucault's observations regarding the concept of 'things' should be noted. Firstly, it is important to note that the focus on things points to a less territory-oriented approach to government. Secondly, the broader understanding of 'things' that Foucault reads into La Perrière's text fits well with Foucault's own ambitions. Foucault observes that 'things' should not be understood in opposition to 'men', but rather as referring to a complex composed of men and things and their subtle and intertwined relations. Furthermore, Foucault observes how such an understanding can readily be related to the often invoked ship metaphor, that is, the metaphor that compares the art of government with the act of governing a ship (135). Even though Foucault delivers this metaphor as a way of presenting La Perrière's thoughts, it can be seen as presenting Foucault's own broader understanding of the art of government as well. In particular, on the backdrop of Foucault's historical enquiries, his concept of governmentality surfaces as the idea that the ultimate end of government is the complicated maintenance of the welfare of the population. Significantly, this idea implies that the population is seen as representing more than the power of a sovereign; that it is seen not merely as the subject of needs, but also as the object utilised for fulfilling these needs (141). Attention to such a utilisation invokes three important and closely related aspects of Foucault's concept of governmentality. Firstly, it invokes the often cited notion of the 'conduct of conduct'. Secondly, it invokes a specific understanding of freedom. Thirdly, it invokes a specific understanding of the relationship between practices and rationalities. These aspects should be considered in slightly more detail.

Foucault continually explores the notion of the 'conduct of conduct', but perhaps he does so most explicitly in the essay "How is Power Exercised?" (1983). Here Foucault observes that the exercise of power should be understood as "a way of acting upon an acting subject or acting subjects by virtue of their acting or being capable of action" (220). The exercise of power is thus characterised as actions upon actions or, as Foucault suggests, 'conduct of conduct'.[1] Foucault further observes that such an understanding of power implies that power is exercised only in the presence of *freedom*, that it is "exercised only over free subjects, and only insofar as they are free" (221). Hence, without freedom one cannot meaningfully talk about power, but rather of mere physical slavery (221–2). Approaching the art of government as a matter of 'conduct of conduct' in which freedom is somehow utilised as being itself a set of powers calls for attention to a whole range of subtle tactics and techniques. Accordingly, the maintenance of the welfare of the population demands a wide range of dispersed governmental forms of action and fields of practices; in Foucault's words, it demands a dispersed 'ensemble' (Foucault, 2007: 144).

Lastly, within such modes of governing, the conduct of conduct is continuously co-constituted by all kinds of counter-conduct. When power is conceptualised as productive, as dispersed at all levels and as efficient to the degree that governors succeed in maintaining and utilising the freedom of the governed to his or her own ends, it naturally follows that power and resistance are inevitably interdependent concepts. Regardless of the apparent logic, Foucault himself only discussed the interdependent relationship between conduct and counter-conduct at any length in one of his lectures. In this lecture, however, he is fairly concise on the matter, providing some illuminative examples of how it is necessary to understand the notion of counter-conduct in order to understand the notion of the conduct of conduct and vice-versa. For instance, he discusses the practice of 'waging war', pointing out how this practice seized to be merely a profession or a general law but became an ethic or a form of political and moral conduct. As being a soldier was no longer just a destiny or a profession, but a

1 Foucault suggests that the equivocal nature of the term 'conduct' is in itself expressive of the 'doubled' understanding of power exercise as a matter of actions upon actions (221).

form of conduct, so another form of desertion practice appeared simultaneously. In Foucault's wording: "[refusing] to be a soldier and to spend some time in this profession and activity, refusing to bear arms, appears as a form of conduct or as a moral counter-conduct, as a refusal of civic education, of society's values, a refusal of a certain obligatory relationship to the nation and the nation's salvation, of the actual political system of the nation" (Foucault, 2007: 265). The example illustrates that counter-conduct inevitably co-constitutes the form of governing that can be characterised as 'conduct of conduct'. When people are governed through their freedom to act, governors constantly have to adapt to the attempts to resist the governance. Or, as Foucault puts it, governors have to rely on the will of the governed ones "not to be governed thusly, like that, by these people, at this price" (Foucault, 2007b, p. 75).

Powers of freedom

Following Foucault, Rose (1999) understands power as a dynamic assemblage of subtle force relations that disperse all levels of contemporary life. Consequently, he talks about 'multiple circuits of power', and he underlines that investigations of governmentality should include attention to the *molecular* level in the aim to "track force relations [...], as they flow through a multitude of human technologies, in all the practices, arenas and spaces where programmes for the administration of others intersect with techniques for the administration of ourselves" (5). As this indicates, Rose's understanding of power as accomplished at all levels implies that power should be investigated at all levels. However, this talk of 'levels' should not invoke the classic micro-macro dichotomy. Rather, when Rose talks about power as dispersed at all levels and the implicated need for small-scale observations, he emphasises that this has to be done in acknowledgement of Latour's insight that the supposedly natural opposition between micro and macro is not so natural after all. More specifically, Rose follows Latour's argument that "the 'macro-actor' is not different in kind from the 'micro-actor', but is merely one who has a longer and more reliable 'chain of command' – that is to say, assembled into longer and more dispersed networks of persons, things and techniques" (5) and, furthermore, that with no differences in actors upon which to found the dichotomy, the dichotomy dissolves. It turns out that it is, after all, not a dichotomy, but rather a web of dispersion. In sum, when Rose talks about the molecular and the 'molar' (6), they should not be considered qualitatively different domains, but rather areas intertwined in differently dispersed networks.

Consequently, Rose argues that studies of power must focus on the numerous endeavours to steer and manage the conduct of others as well as, importantly, the conduct of oneself, and he specifies that studies of governmentality must focus on "the invention, contestation, operationalization and transformation of more or less rationalized schemes, programmes, techniques and devices which seek to shape conduct so as to achieve certain ends" (3). Moreover, he emphasises that while government as simple domination implies the ruining of the 'capacity to act' of the dominated (4), government as the conduct of conduct is the result of the acknowledgement and appropriation of the capacity to act. As he rather blatantly puts it, "[When] it comes to governing human beings, to govern is to presuppose the freedom of the governed. To govern humans is not to crush their capacity to act, but to acknowledge it and to utilize it for one's own objectives" (4). Such governing requires specific forms of resources. Whereas governing through manifest exercises of physical power demands extended material resources, governing human beings through their capacity to act demands an ongoing rationalisation. Contemporary governments need to rationalise their own authority by subscribing to a certain form of truth (27), and, accordingly, the notion of government as the conduct of conduct invokes a specific understanding of rationality that should be considered in more detail.

Rose emphasises that his use of the term 'rationality' neither implies a form of essentialism nor an idea of governing as a predominantly consensus-seeking activity. In particular, Rose recognises that everyday games within the fields of politics are all about compromises 'in the realm of the pragmatics of the possible' and even about simple corruption. Rather, by arguing that governing is in need of rationalisation in accordance with a value of truth, Rose stresses that political games need to be organised within recognisable domains of truth and seemingly coherent systems of thoughts. After all, even the harshest political disagreement has to unfold within the same value of truth to be recognised as a *dis*agreement.

Drawing on Rose and Miller (1992), Rose points out that even though political rationalities are neither very systematic nor closed, they are characterised by certain features. Hence, such rationalities are morally coloured, grounded in knowledge and made thinkable through language (Rose and Miller, 1992: 178–9). In slightly more detail, Rose observes that rationalities have a certain moral form, seeing as they embody the ideals and principles that guide the exercise of government. Furthermore, by pointing out that any given rationality has to be articulated in relation to some conception of the nature of the objects governed, whether these constitute society, the nation, the population or the economy, Rose observes the epistemological character of such rationalities (Rose, 1999: 26–7). Lastly, Rose underlines how political rationalities inevitably invoke a need for a common language. (28)

This understanding of rationality as an inevitable part of governing, bound up with morality, epistemology and discourse, is of particular relevance in this context. Since this book focuses on the intersection of discourse and governmentality, the starting point is to see discourse not only as an aspect of the arts of government, but also as a pathway to an analytics of the rationality and morality of those arts. Consequently, I focus on such matters throughout this part, both in the present chapter on studies of governmentality and in the next chapter on ethnomethodology. Necessarily, then, the relationship between rationality/intelligibility, morality and discourse and the question of how those features co-constitute governing will be touched upon from different, but complementary theoretical perspectives. As I go along, I provide brief indications of how these perspectives can be connected, but I postpone my more detailed discussion of how this can be done and the analytical benefits thereof to the chapter on methodology. For now, Rose's understanding of how *discourse* plays into the dispersed art of governing should be considered in more detail.

As indicated, Rose suggests that an analytics of government implies a focus on the eternal re-rationalisations in the *discursive* fields, that is, in the common, intelligible languages and logics that come off as supposedly common ground. Generally, governmentality studies assume that discursively accomplished *regimes of rationalities* are an inevitable part of governing (cf. Rose, 1999; Miller and Rose, 2008; Dean, 2010), and Rose as well as Rose and Miller point out that language is not mere rhetoric (Rose, 1999: 27; Rose and Miller, 1992: 177). In Rose's words, "Language is not secondary to government; it is constitutive of it. Language not only makes acts of government describable; it also makes them possible" (28). Hence, Rose approaches government and language as an intertwined assemblage that renders it futile to attempt to study one dimension without considering the other. Against this, Rose carefully points out that when he speaks about language, it should invoke neither a static nor a dynamic web of meaningful referents, but the notion of regimes of truth (29). Rose understands 'common language' and 'regime of truth' and 'domain of knowledge' as interchangeable concepts. Furthermore, he underlines that despite the commonality, such domains are characterised by ongoing negotiations of who can say what, when they can say it and for what practical purposes. Fields of language, of truth or of knowledge, are indeed agonistic battlefields, and Rose emphasises that this view of language as epistemological battlefields is differentiated from a view of language as merely a matter of referential meaning attribution, stating "It is not so much a question of what a word or a text 'means' [but] a matter of analysing what counts as truth, who has the power to define truth, the role of different authorities of truth, and the epistemological, institutional and technical conditions for the production and circulation of truths" (29–30). The quote anticipates the later discussion of how an analytics of government can be connected with an ethnomethodologically informed discourse analysis. In particular, the quote stresses an interest not so much in 'pure' or interchangeable meanings, but rather an interest in which 'truths' language affords: who can and cannot say what with which effects at which times and in which places. This interest is readily connected to an ethnomethodological interest in what members in discursive interaction observably accomplish and for what practical purposes. In slightly more detail, exactly like Rose, ethnomethodology focuses on how everyday life is accomplished in discursive interaction. It could be said that when ethnomethodology focuses on members' reflexive use of categories and category predicates in the maintenance of everyday life, it focuses on the ongoing accomplishment of in situ regimes of truth. However, now it is time to consider Rose's discussion of the resource-demanding process of rationalisation.

Rose observes that the inevitable call for rationalisation within multiple and dispersed circuits of power invokes the question: how is governing at all possible? Accordingly, he sets out to discuss the relation between, on the one hand, the myriad of microsites where authorities of all sorts exercise their powers over the conduct of others and, on the other hand, the macro practices of government. More specifically, he notes that this important relationship relies on a complex dynamics which he terms *translation* (48), once again tying back to Foucault. That is, it should be recalled that Foucault points out that the population has to be seen not merely as the subject of needs, but also as the object utilised for fulfilling these needs. When Rose notes that alignments are forged between the objectives of authorities and the projects of those who are the subjects of government, he can be said to be providing a name for Foucault's thoughts on the process of utilisation. In doing so, Rose draws on Callon and Latour. As mentioned, Rose generally follows Latourian insights and dissolves the all too often assumed micro-macro dichotomy for the benefits of a more fruitful *web* of dispersion. Similarly, Rose borrows the term *translation* as well as the related notion of *government at a distance* not only to invoke a general understanding of dispersed assemblages, but also to point out a particularly important element within liberal approaches to government.

Rose's point is not that translation is a phenomenon which is relevant only within liberal mentalities of government, but rather that it is of particular significance within such modes. As I discuss below, this is due to a special ambition within such modes to utilise the autonomy of the subjects of government in the process of the conduct of conduct. Translation becomes important because of the ambition to limit the scope of political activity and, at the same time, heighten the degree of individual autonomy. Consequently, such forms of governing rely on the 'self-organising' capacities within and between individuals to guarantee their well-being, which is, after all, essential for the maintenance of stable government. Hence, for governing to appear to simultaneously respect and promote individual autonomy and guarantee individual well-being, it has no choice but to rely on government at a distance and, by implication, on a smooth translation process. It is, however, no easy task for governors to ensure such a smooth process. Rather, it demands fine-grained and resilient strategies which link the concerns of broader dispersed political rationalities with more narrowly dispersed programmes and problematics within everyday life. Consequently, the translation processes constituting governing at a distance can be seen as the eternal negotiations in which a variety of agents "seek to enhance their powers by 'translating' the resources provided [...] so that they may function to their own advantage" (50). Hence, governing at a distance is only possible "when each can translate the values of others into its own terms, such that they provide norms and standards for their own ambitions, judgements and conduct" (50).

As already indicated, Rose's notion of *freedom* runs through his entire work on current mentalities of government, and in the rest of this section I will consider it in more detail. I begin with a presentation of the main idea of a contemporary utilisation of freedom and proceed with a consideration of how this form of utilisation is related to the current diagram of governing, namely *advanced liberalism*. As part of these considerations, it also becomes clear what the notion of freedom means for the understanding of citizens and citizenship.

Rose's main work on the notion of freedom bears the title *The Power of Freedom*. This title may invoke a number of connotations. Since the value of freedom is currently the guiding principle of lives and dreams, the concept of *freedom* may very well invoke an idea of something inherently good, and so may, by simple implication, the notion of the *powers of freedom*. Hence, the title may invoke the idea that Rose's work is all about *negative* freedom, freedom from interference, or it may invoke the idea that the work is about emancipation and deliberation of the mind. In each case, the concept of power would then invoke one or other idea of empowerment, or one or other idea of the power and desirability of freedom. Whether or not Rose wants his title to challenge such dominating conceptions, it may nevertheless do so. Exactly because the title triggers an unquestionable normativity, it demonstrates perhaps the most important point of Rose's approach to freedom, namely that because we are subjected to living in 'the truth', we are subjected to thinking about freedom in a specific way. In other words, by invoking an idea about powers of freedom as something unambiguously and inherently good, the title demonstrates how freedom is itself a set of powers which is neither good nor bad, but which can be *utilised* for all sorts of purposes. Accordingly, to come to an understanding of Rose's approach to freedom, it is necessary to start to let go of what is thought to be known about the meanings and doings of what is, after all, a fairly elusive concept.

Besides the perhaps ambiguous title, Rose cannot be accused of concealing his ambitions of challenging our commonplace ideas about the concept of freedom. In contrast, he begins his chapter on freedom with the rather bleak observation that not only have the celebrated notions of 'the free world' and 'the free market' succeeded within contemporary liberal politics, they have also infused what used to be radical political thinking (61). Rose does not claim that it is impossible to find differences within the spectrum of contemporary Western politics. He admits that it is possible to point out a range of disputes concerning the explicit definition of freedom and the relative priority of the concept as opposed to other political goals. Rose's critical observation is merely that there is a founding assumption which is never questioned, namely the assumption "that human beings are, in their nature, actually, potentially, ideally, subjects of freedom and hence that they must be governed, and must govern themselves, as such" (62). Consequently, it is this assumption that he sets out to scrutinise. More specifically, it is Rose's ambition to question the unquestionable meaning of the concept of freedom through an 'untimely' diagnosis of the concept.

Rose starts out with two guiding questions: "Could we regard the birth of freedom as an achievement of government? In which case, what would it mean to contest government in the name of freedom?" (62). In particular, Rose is motivated by the understanding that contemporary liberal politics are structured around an opposition between government and freedom. He believes such politics to prosper on their very success in standing out by guaranteeing the highest degree of individual liberty. However, it is not Rose's project to reveal that freedom is in fact a mere illusion. Rather, he follows Foucault and approaches freedom as not only 'real', but as in fact constitutional of government; that is, of the exercise of power. Thus, Rose's project is to investigate the historical opportunities of the existence of specifically this notion of freedom which is compatible with contemporary liberal arts of rule.

In this context, a few overall points on Rose's thorough historical reading will suffice. Rose begins with the observation that even though ideals and practices of freedom have a long history, most authorities agree that it is only within recent Western history that values of freedom have become superior to other values. Hence, the superior notion of freedom that we worship today only became an ideal through the rise of modern individualism and capitalism in the late nineteenth and early twentieth centuries. Specifically, the rise and maintenance of 'the market' depend on an exclusive notion of freedom where persons are accomplished as isolated buyers and sellers of goods (66). This settled, Rose meticulously outlines how this young, westernised concept of freedom has developed from being accomplished as primarily a matter of discipline to being accomplished as primarily a matter of solidarity until, in contemporary societies, it is accomplished as primarily a matter of autonomy. Within the period of freedom as discipline, the liberal strategy of government was to inculcate a strong sense of norms, rationality and private responsibility in the subjects of government (74). However, such an approach could not handle the unintended sides of social fragmentation (such as rising crime), and it became necessary to utilise freedom as a form of solidarity (78). More specifically, it became necessary to accomplish the normal citizen as the social citizen: "Private individuals were simultaneously assigned their social duties [...] and educated in the fact that they must be educated by experts in order to responsibly assume their freedom within the context of society" (83). However, since the middle of the twentieth century, this accomplishment has started to become less productive and has had to give way to yet a new way to act upon human beings, namely a way in which freedom is utilised as autonomy (84). Since this third accomplishment is of particular relevance for my purposes, I will consider it in slightly more detail.

The notion of freedom as autonomy is closely related to the notion of *advanced liberalism*. Rose sees the idea of freedom as autonomy as a technology called upon in the 'dusk' of the social (83). He remarks that within the second half of the twentieth century, the social state has experienced a series of challenges, such as destroyed communities, failure to redistribute wealth and increased interference by authorities and experts (83). He further remarks that as a response to these challenges, a new diagram of governance has taken shape: the diagram he terms advanced liberalism (139–40). This diagram is characterised by the fact that all sorts of social undertakings are now reframed in terms of how they contribute to eternal growth in human capital (142), and, not surprisingly, the diagram is intertwined with a specific conception of the human actor as an *entrepreneur of his or her self*. In this way, as Rose proposes, advanced liberalism is related to a conception of freedom as autonomy (142). Advanced liberalism and freedom as autonomy are accomplished through a form of governing aimed at autonomous individuals in their capacity to realise all of their secular desires; that is, their capacity to live fulfilling lives through consuming acts of choice (84). Within advanced liberalism, the power of freedom is the *obligation* to be free, the obligation to make appropriate, responsible choices and the obligation to actualise oneself. As Rose also puts it, freedom today is an "imperative of self-realisation" (87). To govern today is to govern through informed and knowledgeable subjects and through an individually infused "logic of (responsible) choice" (88).

One last aspect of Rose's account of freedom within contemporary Western ways of governing should be considered. Following Foucault, Rose underlines that the understanding of freedom as infused with power and, as such, as a technical matter that entails different forms of subjectification does not imply that freedom is a sham or an illusion (94). Rose underlines that he understands freedom as the name "we give today to a kind of power one brings to bear upon oneself, and a mode of bringing a kind of power to bear upon others" (96). Hence, Rose's starting point is that human beings are observably able to act upon themselves, and, importantly, he suggests that this is enough to form the basis for minimal normativity; people may, after all, be able to evaluate one or the other practice of freedom in terms of how those who are caught up in it are thereby afforded the resources to sustain or transform the very practice that subjectifies them (97). Following this, Rose suggests that his notion of freedom could in fact generate studies that would be critical in their 'untimely' attempt to evaluate the technologies of freedom in terms of how those who are caught up in them are thereby afforded the resources to sustain or transform the very practice that subjectifies them.

Morality, rationality and regimes of practices

Even though Rose's work on studies of governmentality represents the core of my approach to such studies, it is useful to draw on other researchers to illuminate some of the points of particular relevance to my project. Hence, in this section I draw on Dean's work to further emphasise how rationality and morality are intertwined co-constituents of government. Furthermore, introducing Dean's notion of 'regimes of practices', I point out that governing rationalities are not only bound up with moral questions, they are also inherently embedded in regimes of practice. Lastly, I point out that Dean underlines that an analytics of government should always grow from *how* questions.

Dean (2010) introduces the basic concepts of studies of governmentality, discussing the notions of 'government' and 'governmentality' both. Beginning with 'government', he states that it can be translated into the 'conduct of conduct', and, thus, to understand the notion of government, it is necessary to understand the notion of the 'conduct of conduct'. Of course, this term has already been discussed both in the section on Foucault and in the section on Rose, but Dean importantly adds to those discussions, clarifying how the notion implies an understanding of government as a moral business. Dean stresses that the ethical or moral sense of the phrase 'conduct of conduct' becomes apparent already if one considers the reflexive verb 'to conduct oneself' (17). This indicates the by now known idea that government is not merely about *ordering* people to do this or that. Rather, it involves an attempt "to deliberate on and to direct human conduct", whether one's own conduct or the conduct of others (18). Hence, government is a rational activity undertaken by a multiplicity of actors. Particularly, Dean explicates the distinctive way in which government can be seen as a *rational* activity, explaining that it is a *moral* activity. He points out that the rational attempt to shape conduct can be seen as linked to moral questions: "If morality is understood as the attempt to make oneself accountable for one's own actions, or as a practice in which human beings take their own conduct to be subject to self-regulation, then government is an intensely moral activity" (18). Not only does the quote clarify the close relationship between rationality, moral accountability and government that Rose also notes, it does it in much the same terms as ethnomethodology. That is, although ethnomethodology does not speak about government as such, it argues that the logical, the normative and the practical need to be brought into the same circle of intelligibility.

Like Rose, Dean notes that an approach to government as the conduct of conduct implies a rather specific notion of freedom. Dean ends his discussion of government by observing that government as the conduct of conduct presupposes not only the primary freedom of those who are governed, but also of those who govern. More specifically, he points out that since the notion of freedom entails the capacities of acting and of freedom, it is clear that governing, ourselves as well as others, is fundamentally an exercise of thinking (24). According to Dean, the 'strange' term 'governmentality' can be understood as the rationalities or mentalities of governing (24). He stresses that within studies of governmentality, *rationality* does not refer to a notion of a hegemonic Reason that directs *the* proper reasoning (24), but any relatively systematic way of thinking.

Furthermore, Dean's development of Foucault's notion of *regimes of practices* stresses this *non*-rationalist focus within governmentality studies. Specifically, as Dean points out how rationalities are intertwined with regimes of practices, he demonstrates the non-rationalist presuppositions of the – perhaps misleading – notion of rationalities. Not only is this important within governmentality studies in general, it is also fundamental to my ambition of connecting an *a*-rationalistic, ethnomethodologically informed discourse analysis with such studies. Dean highlights that whereas sociological approaches to collective mentalities have traditionally been oriented in line with collective ways of thinking as relatively easily defined via examination of their social, political and economic conditions, studies of governmentality are interested in how thinking co-accomplishes our systematic regimes of practices (27). Consequently, such studies focus on how thought becomes part and parcel of directing conduct: "The analysis of government is concerned with thought as it becomes linked to and is embedded in technical means for the shaping and reshaping of conduct and in practices and institutions. Thus to analyse mentalities of government is to analyse thought made practical and technical" (27).

Hence, a regime of practices is an assemblage of people going about doing things commonsensically. For instance, it may be possible to consider the commonsensical way we currently go about everyday mobilities to be displaying a regime of *automobility*. Furthermore, it should be noted that as Dean speaks about regimes of practices as "fairly coherent sets of ways of going about doing things" (31), he speaks about them in much the same way as he – and Rose, for that matter – speaks about *rationalities*. Only, when Dean speaks about regimes of *practices*, he speaks about *doing*, not *thinking*. It should be noted that Dean mentions two aspects. Firstly, he points out that whereas regimes of practices are inextricably intertwined with or bound by mentalities of rule or regimes of rationalities, they are not exhausted by the intelligibility of these, and, secondly, he indicates that the relationship must also be acknowledged the other way around. Accordingly, Dean underlines that "the same activity can be regarded as a different form of practice depending on the mentalities that invest it" (26). As I read this, it suggests a more overall point: regimes of rationalities (that is, webs of thinking) and regimes of practices (that is, webs of doing) co-constitute each other and can only be differentiated analytically.

At this point, I would like to sum up two arguments. Firstly, whereas Dean primarily focuses on the importance of not conflating practices with rationalities, his argument implies that it is, in practice, just as futile to speak about rationalities outside the practices they co-constitute. However, secondly, his argument suggests that the relationship of the two types of regimes is not 'equal'. More specifically, Dean considers *practices* the more overall type of regime or assemblage and rationalities only *one* of the co-constituting parts of this assemblage. Accordingly, Dean observes that while practices exist in the milieu of thought, they have a material and institutional locale as well (32). Correspondingly, my point is not that rationalities and practices are coinciding phenomena, but rather that Dean's use of Foucault's notion of regimes of practices demonstrates that rationalities are always immersed in practices; that rationalities are always rationalities within practices.

Lastly, Dean observes that an analytics of government should focus on *how* governing is accomplished. Emphasising *how* questions, an analytics of government attempts to escape any deductive reductionism and to reject any *a priori* distribution and division of power (33 and 37). For instance, approaching automobility as being a regime of automobility would imply bracketing all presuppositions that reduce automobility to a result of certain institutional formations, certain structures of more socio-economic or infrastructural character, or certain ideologies *a priori* assigning more power to a male than a female driver, for instance. On the other hand, automobility would be approached as being composed of heterogeneous elements with diverse trajectories. Whereas none of this can be assumed *a priori*, it can be diagnosed through an analysis that focuses on *how* elements and, by implication, the regime as such are accomplished in situ.[2]

Discontinuities, contingencies and resistance

A point has been made that studies of governmentality have an unfortunate tendency to be 'repetitive' and homogenising, arguing that future studies must focus more on struggle and resistance (e.g. Biebricher, 2008; Bröckling et al. 2011; Malpas and Wickham, 1995). For instance, Bröckling et al. (2011) point out how researchers within the field too often presuppose their own findings, thereby merely duplicating neo-liberal governmentality (16–7). Furthermore, they observe that many studies are characterised "by a closure and clarity suppressing the simultaneous effects of various governmental lines of force" (16). They argue that there seems to be an unfortunate tendency of finalism in the many accounts in which the history of governmentality is presented as "an increasingly fateful correlation of mobilization and self-mobilization, optimization and self-optimization, control and self-control" (17). Against such observations, they observe that governmental programmes of administration are too often treated as blueprints, thus overlooking what Foucault already noted, namely the various forms of resistance. Furthermore, they suggest that this problematic is due to a specific 'methodological quandary' that is insensitive to the contingent and unpredictable forms of resistance and counter-conduct, pointing out that whereas there "is a science of government […] there cannot be one of the art of not being governed" (17). Hence, Bröckling et al. suggest that part of the problem is that resistance withdraws from most attempts to make government available for enquiry, and, drawing on Jacques Rancière (1998), they suggest that within governmental

2 In the next chapter, I demonstrate how this is concurrent with an ethnomethodological starting point. Anticipating things slightly, Garfinkel (1967) critiques what he refers to as a 'favoured' solution: "A favored solution is to portray what the member's actions will have come to by using the stable structures – *i.e.*, what they *came* to – as a point of theoretical departure from which to portray the necessary character of the pathways whereby the end result is assembled" (68).

regimes science itself belongs to the order of police. However, their point is not that studying resistance represents an insurmountable methodological challenge; rather, they suggest that researchers need to be more careful so that they do not end up simply doing *governmentality* in and through doing *research*. In more detail, they argue that it is possible to do a more subtle form of critique by focusing more on *problematisation*. Instead of aiming at a more straightforward or bombastic critique which remains blind to ambivalence, studies of governmentality should aim at "making a performative relation visible in which governmental strategies and patterns of resistance encounter and define each other" (18).

Hence, following Foucault, Bröckling et al. understand resistance as a constitutive moment of governing, that is, they emphasise that it does not make sense to speak about governmental strategies without speaking about resistance, since the two are co-constituting phenomena. This implies that the relationship between governmental strategies and resistance is specifically *contradictory*, "resistance marking both a boundary and a constitutive moment of government" (19). In more detail, Bröckling et al. suggest that government is a necessarily *precarious* affair that is continually confronted with a threefold form of resistance, involving resistance as the *given* (that is, the practice of persisting), the *relinquished* (that is, the practice of changing through negotiation) and the *withdrawn* (that is, the practice or force of the unconscious, manifested through, for instance, repetition compulsion). Therefore, government must continuously readapt and reinvent itself according to inevitable and unforeseen crises of governability (19). Hence, it is Bröckling et al.'s point that studies of governmentality have not paid sufficient attention to the phenomenon of resistance and to the ruptures and disruptions it entails. However, they do not deem the field of studies of governmentality infertile, they emphasise the need for a heightened awareness on matters of counter-conduct and resistance.

Malpas and Wickham (1995) usefully discuss the problem as related to a more general tendency within sociology to focus on success, not on failure. They argue that sociology is imbued with a belief in the possibility of continuous and increasing success, presupposing an understanding of the social world as being, by and large, steadily and unproblematically progressing. Hence, traditional sociology treats failures as temporary breakdowns to be overcome as quickly as possible. However, Malpas and Wickham seriously question such a concentration on success and conceptualisation of failure, questioning whether failure can in fact ever be *overcome* (38). More specifically, they provide a philosophical argument, suggesting that failure is a ubiquitous and central feature of social life, thereby problematising the epistemological confidence in success within most contemporary sociology (45–9). Furthermore, and of more direct relevance, they suggest that it may be possible to attain a more adequate understanding of failure by drawing on Foucault's concept of governmentality (40–5).

Proposing such a Foucauldian sociology of failure, Malpas and Wickham begin by defining governance broadly as "any attempt to control or manage a known object" (40). Furthermore, they observe that such an understanding implies that social life persists in ongoing attempts to control all sorts of objects. They emphasise that these attempts are bound to eternal failure, that there is no such thing as absolute control or, as they put it, that "governance is necessarily incomplete and as a necessary consequence must always fail" (40).[3] Drawing on Latour, they additionally remark that an approach to failure should be oriented to the little details of complex assemblages which "are always implicated within larger assemblages and whose components may operate together in only a loosely integrated, and not entirely consistent, fashion" (42). Against the notion of assemblages, Malpas and Wickham suggest that metaphors of the machine can clarify their argument. They suggest that engines and systems of governance can be compared, in that both are driven by power and employable for powering other systems (42); that is, they can be compared because both are embedded in multiple circuits of power. This not only implies a notion of power as productive, it also clarifies the difference between a Foucauldian understanding of power, only denoting the assemblages of governance, and a more conspiratorial understanding of power à la Gramsci or Althusser (42). More importantly, however, Malpas and Wickham suggest that the metaphor essentially implies that resistance is a constitutive part of governing.[4] Just as a machine needs a specific form of resistance to run, so, too, does government: "A machine is precisely a set of parts that operate together through being placed in such a way that they mutually limit and direct each other. Resistance is not merely the counterstroke to power, it is also that which directs and shapes power" (43). Notably, Malpas and Wickham see this notion of resistance as a certain manifestation of failure. By implication, failure comes off as a constitutive moment of society and it becomes clear that society, essentially, only works imperfectly (43). In line with this argument, I would like to add that if the

3 This conceptualisation is almost identical to Lemke's notion of governmentality as a constant failure of programmes. Whereas Lemke (2002) does not refer to Malpas and Wickham in his later discussion (of course, it could simply be that he has not come across their paper), Bröckling et al. (2011, i.e. Bröckling, Krasmann and *Lemke*) do refer to Malpas and Wickham's paper as a specific point of reference as regards their conceptualisation of resistance.

4 In fact, they remark that Foucault himself used such a metaphor with regard to resistance, namely when he talked about counter-conduct as the 'counterstroke' to power (42).

constitutive aspect of resistance is not acknowledged, it is not at all possible to study governmentalities, but only one's own presuppositions.

The insights from studies of governmentality have certain methodological consequences for this book. Considering both the call to pay attention to the *hows* of discursive interaction at the molecular level and the warning not to disregard the constitutive role of resistance, I find it necessary to take a different methodological approach than that of most studies of governmentality. Studies of governmentality pay close attention to how concepts such as 'freedom' (Rose, 1996), 'dependency' and 'empowerment' (Dean, 2010) are not simply condensed meanings which the analyst has to unmask in order to reveal their ideological content but rather concepts that make certain things and truths possible (79). However, I would suggest that when studies of governmentality conduct analyses with the specific aim of tracing the *hows* of any *particular* concept, they risk presupposing the very importance of this concept. Accordingly, I consider it fruitful to connect studies of governmentality with a method that can make rationalities of government available for enquiry without presupposing the importance of particular concepts, but rather paying close attention to how the very relevance of one concept or the other is also accomplished in and through resistance. Furthermore, I would like to suggest that ethnomethodology can provide such a method and, at the same time, contribute to a more detailed understanding of the important notion of rationality. That is, I suggest that it is possible to draw on ethnomethodology to provide a first stage analysis that can then be interpreted in terms of governmentality. However, to ensure that it can be theoretically sustainable to do so, it is first necessary to discuss the theoretical foundations of ethnomethodology and present a line of ethnomethodological research that can be connected with studies of governmentality more readily.

3. Ethnomethodology

The overall purpose of this chapter is to clarify what I mean when I talk about rationality and morality and to discuss how these phenomena can be made available for enquiry. In more detail, the purpose of this chapter is to draw on ethnomethodology to demonstrate how rationality and morality are continuously negotiated in talk-in-interaction, thereby further illuminating why the study of discursive interaction needs to be included in studies of governance.

I would like to begin by pointing to two overall, somewhat related issues. Firstly, the notion of an ethnomethodological research perspective stems from Garfinkel and his 1967 publication *Studies in Ethnomethodology*. Since the publication of *Studies*, ethnomethodology has developed and come to dominate two research fields. Conversation analysis is concerned with the sequential organisation of talk in everyday and institutional settings, while membership categorisation analysis is concerned with the social organisation of knowledge in categories. Both conversation analysis and membership categorisation analysis stem from the work of Sacks, but the relationship between conversation analysis and membership categorisation analysis, and the relation of these two strands to Garfinkel's ethnomethodology, has been the object of considerable debate. In particular, the debate is characterised by two somewhat demarcated positions, the first holding that Sacks's work represents a more autonomous field of conversation analysis that might be inspired by, but needs to be separated from, Garfinkel's thinking (for instance, Schegloff 1992: xviii); the other holding that Sacks's work is founded in and, importantly, further develops aspects of Garfinkel's thinking (for instance, Button, 1999; Hester and Eglin, 1997; Lynch and Bogen, 1994; Turner, 1974). I discuss this debate in more detail later in the chapter, but I would at this point like to stress that I approach Sacks's work as being thoroughly founded in Garfinkel's work. Relatedly, the work I discuss in my outline of more recent ethnomethodological research carries on the important insights of Garfinkel's founding work. However, since the position holding that Sacks's work initiates a more independent line of conversation analytic research is by far the most dominant, my position is not uncontested.

Secondly, I would like to emphasise an important point, namely that ethnomethodology plays a dual role in this study, representing a theoretical as well as a methodological starting point. The methodological role is, of course, not really observable prior to the analysis, but already in this chapter it should be observed that ethnomethodology itself affords a double role. As demonstrated below, theory and method are interdependent phenomena, and method is not in any way secondary to theory. Whereas the divide between theory, method and methodology is always somewhat artificial, it is almost impossible to keep the concepts nice and tidy within ethnomethodology. Button (1991) also observes that ethnomethodology is never concerned with the 'pure' or isolated theorising of foundational issues; it is more interested in making such issues investigable and available for enquiry (4–5). This is no less true of Sacks than of Garfinkel, and it is against such an understanding that Sacks sets out to extend Garfinkel's notions by applying them systematically to the study of naturally occurring talk. Accordingly, the chapter does not concern pure theorising, but rather theorising through the question of how to enquire into such matters through talk-in-interaction. Hence, whereas the chapter invokes much the same understanding of molecular rationalities and of morality as attempts to make oneself accountable for one's own actions as those figuring within studies of governmentality, it does so by demonstrating how such phenomena are available for enquiry through talk-in-interaction. Consequently, even though the methodological 'manual' on how to utilise ethnomethodology as a pathway for studying rationalities of government is not unfolded until later, much will already surface more and less implicitly in this chapter.

Garfinkel's studies in ethnomethodology

The purpose of this section is to found my approach to ethnomethodologically informed research. On the most overall level, ethnomethodology is founded on the assumption that the meaningful order of everyday life is continuously accomplished. Accordingly, Garfinkel's main ambition is to describe the 'ethno methods' via which this is done. Against this, Rawls (2002) is quite to the point when she argues that ethnomethodology represents the fairly simple idea that if social order is continuously accomplished, then people must have some methods for doing so (5). It might, however, be a little less simple to understand what this idea means and what consequences it has. Hence, to attain a more thorough understanding of the ethnomethodological gist, I will begin this section by considering one of Garfinkel's (1967) so-called *breaching experiments*. Garfinkel designed these experiments to disrupt and, by implication, expose the basic principles of social ordering. Although they were not experimental in any strict sense, it was Garfinkel's ambition that they would be stimulating "aids to a sluggish imagination" (38).

In this experiment, Garfinkel asked a number of students to participate in what was presented as research on new methods of psychotherapy. Each student was asked to consider a personal problem he or she wanted advice on. Furthermore, the student was told that he or she would then be advised at a very high level, but in the form of answers of yes or no to the student's ten questions. The student was told that the adviser would not be in the same room as the student, but in an adjacent room connected by intercom, and the student was asked to disconnect the intercom after every answer and record his or her comments on the exchange. What the student was not told was that the specific sequence of the adviser's answers had been established in advance. Hence, the answers given to the student were not 'real' answers to the student's questions. In reduced form, this is a reproduction of one of the exchanges:

SUBJECT: Ok, this is the situation that I am presented with. I happen to be of the Jewish faith and I have been dating a Gentile girl now for about two months. My dad is not directly opposed to this situation, but I feel at the same time that he is not exactly pleased with it. Mother feels that as long as Dad is not directly opposed to the situation that I should go ahead and continue dating until he makes some direct statement to the contrary. My reason for feeling why he is not too pleased with this is that he has never said don't date her, but at the same time he will come up with digs and sayings that make me feel very ill at ease about dating the girl. My question is, do you feel under the present circumstances that I should continue or stop dating this girl? Let me put that in a positive way. Do you feel that I should continue dating this girl?

EXPERIMENTER: My answer is no.

SUBJECT: No. Well, that is kind of interesting. I kinda feel that there is really no great animosity between Dad and I but, well, perhaps he feels that greater dislike will grow out of this. I suppose or maybe it is easier for an outsider to see certain things that I am blind to at this moment.
I would like to ask my second question now.

EXPERIMENTER: Okay.

SUBJECT: Do you feel that I should have a further discussion with Dad about this situation or not? Should I have further discussion with Dad over this subject about dating the Gentile girl?

EXPERIMENTER: My answer is yes.

SUBJECT: Well I feel that is reasonable but I really don't know what to say to him. I mean he seems to be not really too understanding. In other words he seems to be afraid really to discuss the situation. I mean at least it appears that way to me so far. But I guess if it is worthwhile to me, if I really want to continue to date her that I will go on and have this discussion with Dad. I really don't know what to say because I mean I am dating her. I am not in love with her or anything but I really never know what is going to come out. I guess we should have a discussion based on what the future possibilities might be and how he would feel about that. He may not be too strongly opposed now because we are only dating, but perhaps he sees future complications that he would really like to get off his chest at the moment. I am ready for my third question now. (80–1)

On an overall level, it should be noted that all experiments were completed. Despite the random answers, all students went through all of their questions, apparently quite willing to invest the amount of work needed to figure out what the adviser 'had in mind' if a given answer made no immediate sense (89). On a more specific level, some interrelated principles of common understanding (or, as Garfinkel otherwise prefers it, social ordering) should be noted.

Firstly, the experiment illuminates the *retrospective* and *prospective* features of common understanding. As in the above example, all exchanges revealed that the students believed the adviser's answers to follow from what had unfolded before and to point to what had yet to unfold. In other words, the interaction was sequentially organised. Secondly, the experiment points to the *provisional* character of understanding. This is observable in the students' willingness to revise their earlier understandings in light of the unfolding sequence, and this willingness regards both revision of the understanding of the problem in general and revision of the understanding of past answers more specifically. Additionally, it is observable in the students' willingness to defer their understanding of a troubling answer to a later point in the conversation. Thirdly, the experiment illuminates the normative character of common understanding. The exchanges revealed how the students applied as well as searched for a 'guiding pattern' during the unfolding exchange. This leads to the general conclusion that the principles are *normatively sanctioned* principles of conversation, and that they have a certain 'seen but unnoticed' character that, on the one hand, provides for the ongoing accomplishment of a common world and, on the other hand, means that departures from them tend to result in situations where the 'departing part' is held morally accountable for the departure (89–94).

Since questions of morally accountable conduct are of particular relevance in the present context, it is useful to further consider the issue against Garfinkel's notion of the 'documentary method of interpretation'. According to this method, the process of interpretation can be understood as the activity of approaching 'appearances' as documenting some underlying 'reality' (78). The above experiment can be seen as a study of the documentary method at work. Specifically, it can be seen as an example of the 'no timeout-ness' of the method. Members have to do 'fact production' in flight in that they have to recourse to the enormous array of possible contextualisations to figure out what actual appearances might be documenting, needing to trust that they are in fact participating in sensible documentation. Furthermore, this mutual trust is morally enforced, and it is treated as morally accountable whenever members depart from what is treated as the normal, mutual accomplishment of reality. Consider the following summary of yet an experiment (Garfinkel, 1963):

> The subject was telling the experimenter, a member of the subject's car pool, about having a flat tire while going to work the previous day.
>
> S: I had a flat tire.
>
> E: What do you mean, you had a flat tire?
>
> She appeared momentarily stunned. Then she answered in a hostile way: 'What do you mean? What do you mean? A flat tire is a flat tire. That is what I meant. Nothing special. What a crazy question!' (221)

It is evident from the result that S takes it that E 'breaches' what S assumed was their common understanding of 'what everybody knows' and that S immediately sanctions this breach. In other words, when E does not comply with 'the expectancies of the attitude of everyday life as a morality', the moral constraints of the mutual trust in sense-making become apparent (Garfinkel, 1967: 50). Hence, everyday life is observably *treated* as profoundly normative and morally sanctionable, and just as there is no time out from the documentary method of interpretation, there is no time out from the normative accountability of one's actions. Morality is not nicely and tidily reproduced *after* a time out in which pre-established norms are considered; rather, the moral order is part and parcel of fact production in flight.

Taking the breaching examples and the remarks on morality as a stepping stone, it is time to consider Garfinkel's understanding of the notions of *rules*, *reflexivity* and *rationality*. The concepts are closely related, but for clarification I attempt to focus primarily on one concept at a time. Considering *reflexivity*, Garfinkel observes that the recognisable sense of utterances depends on the socially organised occasions of their use. This is, for instance, observable in the above breaching example where the adviser's yes and no answers were presented sequentially as meaningful answers within a recognisably organised counselling world. Furthermore, the student's persistent, remarkably indefatigable work in this regard resembles Garfinkel's notion of reflexivity: his idea that sense and order are co-accomplished in a continuous process in which members are reflexively oriented to each other and, importantly, to the co-accomplished context.

This understanding of reflexivity informs the approach to norms or *rules*. Observing that the activities whereby members produce and manage settings of organised everyday affairs are identical to members' procedures for making those settings accountable, Garfinkel suggests that rules are always applied 'for another first time'. It must, after all, require some sort of 'seen but unnoticed' background expectancy, if members should be able to simultaneously accomplish and recognise the *moral order* of everyday life.[1] However, Garfinkel addresses his understanding of rules (more) explicitly when he discusses the difference between members' actual occasions of interaction, and members who are doing sociology. Whereas the former type of accomplishment is omnipresent, unproblematic and commonplace, the latter is somehow required to treat the rational properties of practical activities as 'anthropologically strange' (9–10). Accordingly, accounts or rules are always used and recognised for another first time. In this way, Garfinkel simultaneously acknowledges the rule-bound orderliness of everyday life and insists on the inherent reflexivity of this ongoing structuring.

Against the intertwined notions of rules and reflexivity, it should not be too surprising that Garfinkel approaches rationality as an inherent feature of the process of order and meaning-making. More specifically, he discusses rationality as straightforwardly observable when conversationalists make use of the practices of 'et cetera', 'unless' and 'let it pass' to demonstrate the rationality of their achievements (Garfinkel, 1967: 3). Rationality, then, is the accomplished demonstration that what is done is consistent and in accordance with common rules, with 'what everybody knows' or with common sense *for all practical purposes*. At this point of reconsidering studies of governmentality, it should be recalled that regimes of rationality refer to whatever domain at a given time and place constitutes the complex field of doable 'true' accounts. A main interest of such studies is to consider the conditions of existence of such fields, especially from a historical perspective. In contrast, the ethnomethodological understanding of rationality would insist that such 'fields' are always accomplished in situ 'for another first time'. However, the approach to rationality as *accomplished*, that is, as contingent rather than necessary, is shared. Furthermore, whereas studies of governmentality tend to focus on the more molar level, Rose, for instance, agrees that it is equally important to study how rationalities are accomplished and for what practical purposes at the molecular level of everyday interaction. Hence, the difference seems to be that whereas studies of governmentality focus on the given conditions of a given regime, ethnomethodology focuses on how a given regime is actually accomplished here and now. However, as I discuss later, this difference may not be a difference of type, but merely of perspective.

As invoked above, Garfinkel's unique focus on social structures as continually accomplished leads him beyond any traditional understanding of methods, and,

---

1 The idea that the background expectancies of everyday life are essentially 'seen but unnoticed' is but one of the notions Garfinkel borrows from Schutz. Garfinkel (1967) remarks that Schutz is one of the only social theorists that pays attention to these important features of daily life (37).

as a consequence, the ethnomethodological approach to method is both famous and infamous. Garfinkel (1967) observes that ethnomethodology, as a study of practical actions, should be conducted in accordance with five policies (31–4). The policies are not very clearly distinguished, and I take the liberty to order, emphasise and omit policies according to my specific purposes. Hence, I discuss the policies via two more general observations.

It should firstly be remarked that Garfinkel emphasises that studies in ethnomethodology cannot and should not disqualify any research domain by reference to predefined principles. On the other hand, such studies should recognise that "any occasion whatsoever" (32) can be the starting point for an enquiry into socially organised, artful practices. Ethnomethodology, in other words, is concerned with the description of *any* method that members draw on in their ongoing production of recognisable social orders (32). Correspondingly, one important ethnomethodological policy is to pay equal interest to all sorts of domains. This is, of course, an inherent consequence of the understanding of reflexivity and the related idea that the activities whereby members produce and manage everyday settings are identical with members' *methods* for doing so in a meaningful and accountable way. In addition to explicating a method policy, the formulation also negatively questions the feasibility of traditional selection criteria: namely how the tendency to prioritise conveniently administrable or politically highly profiled sites can at all be sustained.

Secondly, Garfinkel's directions on how (not) to study rationality should be mentioned. Again, Garfinkel's policy evolves from the same foundational approach to practical actions and order as ongoing and reflexively accomplished. Accordingly, his policy is based on a rejection of any method that in one way or the other is oriented to the rational properties of members' actions as matters of members' compliance to pre-established rules. Against this, Garfinkel argues that all sorts of rational properties should be methodologically approached as contingent achievements of organisations of common practice: "These phenomena are contingent achievements of organizations of common practices, and as contingent achievements they are variously available to members as norms, tasks, troubles" (33). This, of course, also implies that rational properties can never be described by using a rule or a standard obtained *outside* the settings concerned. Or, more explicitly, it is in principle impossible to presuppose rational properties, and studies of ethnomethodology should therefore not decline to such speculations but rather pay close attention to whatever properties are demonstrably recognised, used, produced and talked about by the members of a given setting. By declaring it the heart of the matter, Garfinkel emphasises the policy as well as his project as such: "The demonstrably rational properties of indexical expressions and indexical actions is an ongoing achievement of the organized activities of everyday life. Here is the heart of the matter" (34).

Concluding the section, consider that Garfinkel's research policies point to his notion of 'the judgemental dope'. The notion of the dope stems from his general observation that social science theorists have in general overlooked that social order is fundamentally reflexively constituted. It is, however, in a discussion of common approaches to standardisation that he introduces the notion of the dope. Garfinkel observes that social science theorists commonly presuppose the very fact of standardisation to explain when actions comply with standardised expectancies. Faced with the problem of how to provide for the ways in which expectations concerning ordinary conduct are routinely fulfilled, social theorists commonly, as a 'favoured' solution, "portray what the member's actions will have come to by using the stable structures – *i.e.*, what they *came* to – as a point of theoretical departure from which to portray the necessary character of the pathways whereby the end result is assembled" (68, emphasis in original). According to Garfinkel, the reflexive constitution of standardisation (that is, the fact that members discover, create and sustain standardisations exactly as they act in accord with them) is thereby neglected, and members are made out to be mere judgemental dopes of a cultural and/or psychological sort (66–8). Hence, the notion of the dope can be seen as Garfinkel's, perhaps slightly sarcastic, way of emphasising what he considers to be the main problem within traditional social science, namely that it implies a world of preconfigured zombies. However, it should be noted that this is not to be understood as first and foremost an *ontological* case against determinism. In particular, acknowledging that ethnomethodology is a fundamentally *a*-ontological position, the somewhat loaded notion of a *judgemental dope* should not invoke the idea that Garfinkel is mainly critiquing that such dopes are, ontologically speaking, deprived of free will. Rather, it should be noted that his critique is prominently *methodological.*

Garfinkel's demonstration of the problem of a world of dopes is of particular relevance to the book, since it, more than anything else, emphasises the methodological importance (or, as it is, necessity) of paying attention to the *hows*. Whereas studies of governmentality generally recommend that the *hows* at the molecular level should not be neglected, nobody illuminates as clearly as Garfinkel that anything else is at best in vain. In particular, Garfinkel makes it clear that if members are made into dopes, the analyst risks presupposing his or her analysis. In a study such as mine, which aims at a first stage analysis that can then be interpreted in terms of governmentality, such presupposing could end up merely duplicating neoliberal governmentality (cf. Bröckling et al., 2011). Considering these severe prospects, the next section considers in more detail *how* to pay proper attention to the molecular *hows*, focusing on the methods founded by Sacks: namely membership categorisation analysis and conversation analysis.

Sacks on sequentiality and categorisation

If it was difficult to define the boundary between theory and method in Garfinkel's work, it is more or less impossible in Sacks's as he sets out to extend Garfinkel's notions by applying them systematically to the study of naturally occurring talk. Consequently, this section necessarily anticipates some of the tools I return to when I present the methodology of the study in Chapter Five.

To introduce Sacks's impressive work on conversational sequence, I begin with a brief introduction to his very first lecture in the fall of 1964, "Rules on conversational sequence". In the lecture, Sacks presents three first exchanges in telephone conversations recorded at an emergency psychiatric hospital. All of the exchanges occur between people who have not talked to each other before, and in all of them 'A' is a staff member of the psychiatric hospital, and 'B' is the person calling either about his or her own psychiatric troubles or about somebody else's problems. The exchanges go like this:

1: A: Hello // B: Hello
2: A: This is Mr Smith may I help you // B: Yes, this is Mr Brown
3: A: This is Mr Smith may I help you. // B: I can't hear you. // A: This is Mr Smith. // B: Smith. (Sacks, 1992a: 3)

These conversational openings lead Sacks to a number of observations. Firstly, he observes that in the first two exchanges, B's answers seem to reflect the format provided by A's first actions: 'hello' follows 'hello', and a name follows a name. Secondly, he observes that by being ordered in such pairs or units, the first turn of the first two examples provides a specific slot for the second. In neither case does the first pair part speaker ask specifically for a 'hello' or for the other's name but just makes it observably relevant (4–6). Sacks's point is *not* that the second pair part speaker has to provide the projected response, but that he or she can in fact utilise the slot for all practical purposes, for instance, for avoiding giving his or her name in response to having been informed of the other's. Seemingly, this is what happens in the third exchange in which the caller's declared problem in hearing is a methodical way to avoid giving his or her name. B does not simply *ignore* what he or she was projected to do, but circumvents it in a much more exquisite way.

Drawing on these three examples, Sacks demonstrates how rules or orders are accomplished in talk-in-interaction. Furthermore, by pointing out how B in the last example carefully uses a device for circumventing the projected response without *disregarding* what he or she ought to do, Sacks demonstrates that this ordering is not a matter of reiterating internalised rules, but of accomplishing them for another first time and for all practical purposes. Accordingly, like Garfinkel, Sacks is not interested in revealing some sort of underlying Reason why members go about producing what they produce. Rather, he is interested in how rationalities are continuously accomplished as part and parcel of the prominent task of doing everyday life. As it will become clear in the two next sections, Sacks thereby initiated a line of work that respecifies or translates what is often thought of as philosophical problems into practical ones.

Similarly, presenting the basic principles of membership categorisation analysis, I focus again on one of Sacks's lectures on conversation (Sacks, 1992a: 236–51). In fact, the lecture is not directly on conversation, but on a short 'story' by a little girl, namely the story "The baby cried. The mommy picked it up". On the basis of his discussion of this example, Sacks derives the basic principles of membership categorisation analysis. The purpose of Sacks's investigation is not the everyday one of 'merely' making sense of a story, but of explicating the sense that has already been made. In other words, Sacks's first step is to explain how he understands the story, and his next step is to design an *apparatus* that can provide for the fact that he and other members come to that particular understanding (244).

The first element in this apparatus is what Sacks calls membership categorisation devices. According to Sacks, a membership categorisation device is a collection of, on the one hand, categories for referring to people and, on the other hand, rules of application (246). Collections are not just any set of categories, but categories that routinely and commonsensically are considered to 'go together'. Accordingly, 'mother' and 'baby' are examples of categories, and 'family' is an example of a membership categorisation device.

Further, Sacks describes two rules of application with regard to membership categorisation devices (244). The first rule, the economy rule, states that while I am often, in one sense, ascribable to more than one category (for instance, 'woman', 'left winger', 'Dane' and 'assistant professor') I am not always 'doing' these categories in equally relevant ways. Which category comes off as relevant in a given situation depends on that, thereby constituted, situation, but ordinarily, *one* category will provide a sufficient description of me in any actual situation. The second rule, the consistency rule, observes that if somebody else has been categorised using the category 'Swede' from the device 'nationalities', then I *should* be categorised using a category from the same device, namely 'Dane'.

Based on the consistency rule, Sacks derives the corollary known as the hearer's maxim which states that if two categories can be heard as categories from the same collection, they should be heard as such (247). This maxim provides for part of the hearing Sacks presented in the introduction to the lecture. The category 'baby' is a part of the membership categorisation device 'family', but, at the same time, it is a part of the membership categorisation device 'stages of life' (and, in fact, any other observably relevant device). However, given that the category 'mommy' is a relevant category in the membership categorisation device 'family', but not (as readily) in the membership categorisation device 'stages of life', the corollary to the consistency rule yields that the relevant membership categorisation device is 'family', or, more specifically, the relevant membership categorisation device is restricted to 'family'.

To provide for the fact that he does not consider the mommy any given mommy, but the child's mommy, Sacks expands the characterisation of the membership categorisation device with what he calls duplicative organisation (247). Duplicative organisation is characteristic of some, but not all, membership categorisation devices, such as the membership categorisation device 'family'. If a membership categorisation device is characterised by duplicative organisation, the individual categories in the membership categorisation device are believed to define a unit. Consequently, if members have been categorised using a category from such a membership categorisation device, they are considered 'cases' of the unit (248).

Sacks adds one more additional observation which is most relevant in the present context, namely the observation that the baby was picked up *because* it cried (244). That is, Sacks explains the apparent causality between the first and second events; that is, between the crying and the picking up, through the notion of category boundedness and category-bound activities. Category-bound activities are activities that are *expectedly* carried out by the member of a given category. Crying, for instance, is expectedly done by members of the membership category 'baby' in the membership categorisation device 'stages of life', and picking up babies is, in turn, expectedly done by members of the category 'mommy' in the membership categorisation device 'family' (249). Importantly, category-bound activities and categories are co-selected. The activity provides for the category, and the category provides for the activity.[2] Based on the concept of category boundedness, Sacks derives yet another maxim that provides for the hearing that the baby belongs to (at least) the membership categorisation

2 The notion of category boundedness has been expanded by others, particularly Watson (1978), and Jayyusi (1984).

device 'stages of life' (250). This, in turn, finally provides for the fact that the mother is the mother of the baby.

Recent debates on sequentiality and categorisation

The purpose of this section is to depict the particular line of ethnomethodological research that I draw on in the book. In the understanding that Sacks carries on Garfinkel's thinking, the section takes its starting point in Sacks's original work in conversation analysis and membership categorisation analysis; however, as indicated in the introduction to the chapter, it is a fundamental assumption of the book that these two lines of research are interdependent. This should be clear from the discussions below, but already now I would like to emphasise my position by naming it 'EMCA'. It is common abbreviation practice to refer to ethnomethodological research that does not draw on conversation analysis or membership categorisation analysis as EM. Similarly, 'pure' conversation analysis is abbreviated CA, and membership categorisation analysis is abbreviated MCA. When researchers want to stress that their conversation analytic approach is based in ethnomethodology, they abbreviate their position EM/CA, whereas an EM/MCA abbreviation is not common. However, seeing as I want to emphasise that I understand conversation analysis and membership categorisation analysis as interdependent and as founded in ethnomethodology, I have chosen the non-standard abbreviation EMCA to characterise my overall approach and at the same time give up all other abbreviations (except, of course, when quoting others). In line with this, the purpose of this section is to outline that a thoroughly ethnomethodological approach to category work can benefit from insights from conversation analysis as well as from insights from studies of governmentality.

Reconsidering the debate on how Sacks's work should be properly received and developed, conversation analytic scholars such as Schegloff have criticised scholars doing membership categorisation analysis for misunderstanding Sacks's original work in this field. According to Schegloff, Sacks did not want to establish membership categorisation analysis as an independent research field. Rather, Sacks just wanted to investigate the observation that certain categories and activities seem to 'go together', and according to Schegloff (2007b) "it took [Sacks's] very special cast of mind to pry [the categorization and person description] out of his data" (481). Therefore, Schegloff holds that others should stay within the field of conversation analysis and approach membership categorisation analysis only where the latter can be explained in terms of the former in terms of a cumulative science of conversation.

In contrast, a second position suggests that Schegloff tends to forget the spirit and breadth of Sacks's work, not least his ethnomethodological heritage and his work on categories. According to Hester and Eglin (1997), Schegloff has pushed the development of conversation analysis in a mechanical direction, where the description of conversational structures is separated from their context of use at the risk of missing the indexicality of meaning and the reflexivity of action and context. Housley and Fitzgerald (2002: 59) make a similar observation, pointing out that the consequence of the massive conversation analytic focus on sequential organisation is that other important dimensions of members' work within interaction have been neglected.[3] Accordingly, this second position is roughly characterised by the understanding that sequence and category should not be considered independent phenomena in talk-in-interaction (Hester and Eglin, 1997: 2). As already stated, I follow this position, understanding category analysis as an inevitable part of ethnomethodological research. Accordingly, throughout the rest of this section I go into more detail on the nuances of the arguments holding that category analysis is somehow important, thereby pointing to my own position.

Watson is presumably the ethnomethodological scholar who has paid most attention to the interrelationship between membership categorisation and sequential elements. Watson (1997) holds that sequentiality and category work are interdependent or, in his own words, that they represent "two sides of the same coin" (73). According to Watson, sequential details alone cannot provide a complete description of members' methods. Consequently, if analysts look only at the sequential organisation of talk-in-interaction, they risk drawing on categorial aspects of conversation 'unknowingly'. Watson remarks that, for instance, in the physical transcripts of recorded interaction, interlocutors are often identified on the basis of their function or position in the institutional setting, such as 'D' and 'P' in doctor-patient interaction and 'M' and 'D' in mother-daughter interaction. However, Watson's point is that such categories are not always made observably relevant in data but rather are selected by the analyst who thereby draws on them as unexplicated interpretative resources in analysis (51–2).

The relationship between membership categorisation analysis and conversation analysis is also debated at present. In more detail, Stokoe (2012) initiates a discussion on the proper understanding and use of membership categorisation analysis based on her observation that the present situation can be described as

3 See also Hester and Francis (2001), Lynch (1993, 2000), Lynch and Bogen (1994, 1996) and Watson (1994, 1997, 2001).

one in which "CA is the 'juggernaut' to MCA's 'milk float'" (278). She notes that membership categorisation analysis risks to die out as a distinct discipline because of Schegloff's and others' positioning of membership categorisation analysis as an area within conversation analysis. As Stokoe notes, this is observable, for instance, when Wilkinson and Kitzinger (2008) talk about how conversation analytic methods can be extended to areas like person reference and membership categorisation, describing it as one of the "most vibrant areas" within conversation analysis (565).

However, Stokoe's point is not that such a development would necessarily be regretful. Rather, she remains "ambivalent with regards to the maintenance of distinct methods versus the study of categories being subsumed into CA, as one of its 'generic orders of conversation'" (278). Hence, whereas her starting point is that it might be useful to pursue the subsumption of membership categorisation analysis into conversation analysis, her point is, nevertheless, that the former can provide important insights into culture-in-action, identity and morality, inference and meaning etc. (278–9; for such studies, see, for instance, Eglin and Hester, 1999; Housley and Fitzgerald, 2007; Plunkett, 2009; Summerfield and McHoul, 2005). In this way, Stokoe's starting point is at least partially aligned with that of Hester and Eglin and Watson, in that it does not treat sequential and categorial concerns as analytical alternatives. However, Stokoe's conclusion seems to be somewhat different from Watson's position. Whereas Watson points to the inevitability of categorisation analysis, Stokoe's point appears to be that if approached in careful accordance with certain principles, categorisation analysis does not necessarily *ruin* the sequential analysis and may even add something to it (Stokoe, 2012: 299).

It is characteristic of most of the commentaries following Stokoe's paper that they aim to provide an assessment of whether her suggested principles can in fact guarantee systematic analyses that escape the dangers of the 'wild and promiscuous'. Hence, Silverman (2012) is aligned with Stokoe's position and finds her principles useful. Gardner (2012) is a bit more sceptical of categorisation analysis, but agrees that Stokoe's method may provide insights beyond common sense. Whitehead (2012) also generally endorses Stokoe's contribution, but he emphasises that the strengths of Stokoe's principles are that they call for actual analysis, not just methodological discussions. Rapley (2012) starts by pointing out that he completely agrees with Stokoe's project, but he nevertheless questions whether her principles can embrace the subtle ambiguities of everyday interaction.

However, Fitzgerald (2012) is more sceptical of the implications of Stokoe's project. Whereas Fitzgerald welcomes the opportunity to discuss the question of the relationship between categorisation analysis and sequential analysis, he argues that Stokoe's paper suggests a not too interesting development of categorisation analysis. He observes that (much) conversation analysis has for some time developed in a less ethnomethodological and more formalistic direction, and that Stokoe's principles suggest that membership categorisation needs to take the same direction (307). However, according to Fitzgerald, it is not at all clear why membership categorisation has to be developed on the terms of more corpus-oriented conversation analysis "as opposed to ethnomethodology's terms, Sacks's terms or members' terms" (307). In fact, Fitzgerald remains sceptical that membership categorisation analysis can at all benefit from engaging with conversation analysis, and he suggests that membership categorisation is likely to benefit from other approaches as well (310). Accordingly, he rejects the premise that categorisation analysis is dying, and he concludes by emphasising both its vitality and its interdisciplinary potentials (310), proposing that not only does categorisation analysis not need to comply with the conversation analytic ideals of a cumulative science, it may even gain more from insights from other disciplinary fields. In this way, Fitzgerald encourages attempts such as mine to connect ethnomethodology and studies of governmentality. Accordingly, my more specific approach to EMCA can be characterised as a combination of Watson's and Fitzgerald's respective arguments.

An endogenous approach to moral order

Having established my overall approach to EMCA, it is time to consider how it can be seen as a way of enquiring into the endogenous, continually accomplished moralities of practice, or as a pathway to the study of the rationalities that co-constitute governing. Hence, it should be recalled that rationality and morality are seen as closely related phenomena both within Garfinkel's ethnomethodology and within studies of governmentality. For instance, both Rose and Dean observed that the rational attempt to shape conduct is inevitably bound up with moral questions. Similarly, Garfinkel does not treat rationality and morality as separate fields of enquiry; rather, he discusses how activities are continually accomplished as morally accountable and rational. Accordingly, whereas this section focuses on the moral features of rationality, it is important, if obvious, to bear in mind that morality is approached as an *intelligible* phenomenon.

Jayyusi (1984; 1991) is without doubt the person who has done the most to develop Garfinkel's observations concerning the moral order of practical action, drawing on and expanding Sacks's work on membership categorisation. In greater detail, Jayyusi (1991) bases her argument on Garfinkel's observation of the normative grounding of social order, emphasising that his experiments demonstrate how moral order is constitutive of intelligible, practical action and understanding (235–7). She then observes that Sacks's work can be said to carry on the implications of Garfinkel's work, pointing out that this can be seen both in his (and Schegloff's) work on sequential organisation and in his explication of the logic of membership categorisation devices (237).

In particular, Jayyusi points out how Sacks's distinction between appropriate and correct description "marks out, more than anything, the moral weave, and the moral groundings of ordinary discourse" (237). For instance, Jayyusi demonstrates how the ordinary version of Sacks's (1992) story 'The baby cried. The mommy picked it up' provides the routine unproblematic hearing that the mother picking up the baby is the mother of the baby. Then she discusses an alternative version of the story, namely 'The baby cried. The scientist picked it up'. In this story, the second categorisation is not readily heard as stemming from the same device as the first, and 'picking up a baby' cannot readily be heard as an activity bound to scientists (238). Hence, in this case it would not be co-operative to hear the scientist as the mother of the baby. Rather, based on a default orientation to the relevance of categorisations, the story would perhaps be heard as contexted in a laboratory setting (238). As Jayyusi remarks, the presumption that the task at hand is relevantly accomplished invokes Garfinkel's idea of trust, and it becomes clear that communicative praxis is inherently moral in that it presupposes an ethical 'stance', a mutual trust. In this way, Jayyusi's work clarifies how an analysis of the conduct of conduct can draw on ethnomethodology. In particular, recalling Deans' observation that the moral side of governing is observable in peoples' accounts for their conduct, it is clear that it can be seen as a matter of governing when one or the other category is accomplished as appropriate for all practical purposes.

Furthermore, Jayyusi discusses how membership categories in and through their reflexively constituted character provide for the moral constitution of praxis. Of particular methodological relevance, Jayyusi observes that Sacks's notion of category-bound predicates importantly provides for the moral accountability of certain actions and omissions (240). Sacks's work, then, not only demonstrates the organisation of members' conventional knowledge of the world; it simultaneously shows that such knowledge is morally constituted and constitutive of moral praxis. Hence, knowledge is constitutive of, and provides for, what Jayyusi calls "a moral inferential logic", that is, "a logic of moral inference that is at the same time a moral grounding of practical inference" (240). According to Jayyusi, everyday category work involves such moral inference work in situ. For instance, whenever members use the mundane reference or description 'mother', it provides for a variety of inferential trajectories that are more or less moral due to the more or less morally implying character of the features. That is, the bound rights and obligations of the category are ordinarily treated as morally implying features, and this is why the 'mother' is readily heard as having both an obligation and a right to pick up the crying baby – a matter that renders her morally accountable if she does not pick it up. It needs to be stressed, however, that this moral constitution of praxis is evidenced in members' actual practices. For example, people may routinely draw on the obligations bound up with their membership of an 'employee' category to resist other obligations (such as, for instance, a possible obligation to drive in a less environmentally harmful way). Hence, the point is not that the category 'employee' or the category 'mother' is bound up with a prefigured and/or universal set of morally implying features (it is not the point that being a mother to a baby inevitably constitutes a moral obligation to pick up the baby when it cries). Rather, when Jayyusi draws on a story in which a mother picks up a crying baby, it is not to claim what mothers per se do and do not do, but to trigger the acknowledgement that (whether we hear the category 'mother' as co-constituted by, for instance, an obligation to pick up her baby) the intelligibility of the story, of any story or piece of interaction, is reflexively constituted in practico-moral terms. In Jayyusi's words: "The practice, in which our category concepts are embedded and used, and the appraisal, the con-ceptual, moral, and practical are reflexively and irremediably bound up with, and embedded in, each other" (241).

Before considering Jayyusi's general points on the implications of her ethnomethodological relocation of the site of enquiry into moral order, it should be noted that Jayyusi mentions one more ethnomethodological domain, focusing on the *hows* of this intertwined relationship of the practical, the moral and the conceptual, namely the domain of conversation analysis. Most prominently, she considers the notion of 'preference' (Pomerantz, 1984). Reiterating the observation that first pair parts of adjacency pairs project a context in which a specific second pair part is the preferred, Jayyusi (1991) reminds us that this 'preference' is not due to a matter of empirical regularity, but of normativity seeing as 'preference' is "grounded in our understanding of the contingencies of practical actions and their embeddedness in a moral context" (242).

Jayyusi concludes her argument for an ethnomethodological relocation of morality, summarising the critical ramifications of it. She does so by highlighting the most important points constituting her notion of a *praxiology of moral order*. Remodulating Jayyusi's list of points slightly, it should firstly be observed that morality has a modal logic. In more detail, Jayyusi points out that just as all other actions are sensitive to *and productive of* various contexts and achievements, so are moral actions. Hence, "even though [conventionality] is what makes possible the production of a shared and known in common social order in the first place,

[it] is nevertheless *praxiologically given to members*" (244). Secondly, it should be acknowledged that moral diversity and difference are given in this logic of conventionality. The logical grammar of concepts never fully determines what can be done. Rather, concepts are, in Sacks's words, bound up with a multiplicity of features, and there is, in Wittgenstein's words, a multiplicity of criteria for their use. By implication, difference is not a disorder, but rather emerges from a specific practical order of reflexively available options made relevant in the conduct of practical life (245). Thirdly, such a praxiological understanding of moral order implies that objectivity is praxiologically constituted. Importantly, however, this does not entail that differences in moral judgement are free-standing or indeterminable. Rather, differences are always presented on a background of common languages and can as such always be recognised and negotiated and assessed *in* these languages (245–6). Lastly, Jayyusi notes that moral values are not private or inaccessible. Significantly, moral values are publicly available as they are displayed and negotiated as an ever-present aspect of actors' actions and discourses (243–4), and, consequently, since members are always already in the midst of (also) moral businesses, there is no exit from the moral order (247).

These points reinvoke Garfinkel's observation that members are ordinarily oriented to daily life as inevitably bound up with morality and his related expression of 'the attitude of daily life as a morality'. When Jayyusi emphasises that there is no exit from the moral order, she is carrying on Garfinkel's observations that everyday life is observably treated as profoundly normative and morally sanctionable and that there is no time out from the normative accountability of one's actions. Jayyusi further considers the consequences of her claim that "one cannot get out of the moral order in order to talk about the moral order" (247), pointing out that the claim has certain implications for the analyst, namely that the analyst can never escape his or her own moral resources, even when he or she turns the moral order into a topic (247). Hence, also concerning the position of the analyst, Jayyusi fully acknowledges the implications of her argument that the conceptual, the practical and the moral are mutually embedded in each other. In fact, she further underlines her position, touching upon the notion of 'ethnomethodological indifference'. Her discussion of this notion clearly outlines not only *her* position in the field of ethnomethodology, but also the position I take.

The notion of 'ethnomethodological indifference' stems from Garfinkel and Sacks (1970) who present it as an ethnomethodological research policy (1970). In more detail, the notion marks the difference between traditional sociological reasoning and ethnomethodology, and in line with both the ethnomethodological interest in ordinary activities and everyday life and the ethnomethodological ambition to abstain from any 'ironising' practices, Garfinkel and Sacks describe the 'indifference' as a procedural rule, holding that ethnomethodological studies aim to describe *any* member's accounts "while abstaining from all judgments of their adequacy, value, importance, necessity, practicality, success, or consequentiality" (345).

Whereas some ethnomethodologists have taken this policy to imply that ethnomethodological description can never be extended into criticism, Jayyusi importantly points out that her work on values and moral judgement suggests that it is unclear why it can be concluded that meticulous analyses of praxis cannot be used as a starting point for "political pragmatics" (Jayyusi, 1991: 247).[4] Jayyusi observes that indifference should be understood much more 'localised', as something which should be obtained "from *within* the analytic descriptivist 'moment'" (emphasis in original). The emphasis on 'within' and the choice of 'moment' should be noted. According to Jayyusi, it needs to be recalled that an indifferent description also constitutes an *irremediably* embedded activity (251, n. 17). That is, she observes that even though the analyst must aim to describe what *members* focus on as observably relevant, he or she of course has to draw on his or her "own membership to identify and describe such relevances in the first place" (251, n. 17). Consequently, the 'descriptive moment' is an ongoing accomplishment and cannot be appointed any independent status, or, as Jayyusi puts it, an indifferent description cannot be extended beyond its moment of "'uncovering' without erasing the 'boundaries' which provided its context and point in the first place" (251, n. 17).

Jayyusi thus suggests that since the ethnomethodologist is always already involved in making moral judgements about the practices available in situ, there is no clear reason for rejecting that ethnomethodological analyses can provide

4 Jayyusi draws on a discussion between McHoul, on the one hand, and Bogen and Lynch, on the other hand, which concerns the question of critique in relation to ethnomethodology. McHoul raised the question of the possibilities of 'political pragmatics' in a paper focusing on the work of Coulter (McHoul, 1988). That paper led to a critical discussion in the *Journal of Pragmatics*: firstly, a response by Bogen and Lynch (Bogen and Lynch, 1990) and, secondly, a reply to the response by McHoul (McHoul, 1990). The discussion concerned McHoul's suggestion that Wittgensteinian-inspired ethnomethodology can be read as a crucial starting point for political pragmatics, and Bogen and Lynch argued against this suggestion. Inspired by Jayyusi's work, McHoul advanced his notion of a critical ethnomethodology in a later paper which I discuss in the next section, summing up this chapter on ethnomethodology with an outline of how the theory can be politicised.

a starting point for political discussions and, furthermore, for rejecting the usefulness of elaborate analysis, focusing on how in situ available options can be treated as political (249). In fact, Jayyusi proposes that rejecting the idea that ethnomethodological analysis can ever be the starting point for critique leads to a paradox: the paradox of defending iterative, value-free descriptivism in the hope of abandoning it. Hence, in the end Jayyusi provides two important, interrelated arguments. Firstly, she emphasises that since any business is always already a moral business, the interactional business of accomplishing a meaningful rational world is always already a business of accomplishing a morally ordered world. Secondly, taking seriously her own argument that interaction is always imbued with morality, Jayyusi observes that it is not necessarily contradictory to be engaged in an ethnomethodological analysis and in a political project at one and the same time. On the contrary, she suggests that some interpretations of ethnomethodological indifference lead to a paradoxical position, arguing for the impossibility of descriptions that are neither postulated to be value-free nor understood as politicised or critical in any understanding of the terms. This latter point is pursued in the next section which concludes the chapter with a discussion of a critical ethnomethodology.

Politicising ethnomethodology

McHoul (1994) pursues Jayyusi's rare argument that studies in ethnomethodology are inherently "politico-moral".[5] Since McHoul's argument draws extensively on Derrida (in particular his *Of Grammatology*, 1976), it is rather specific, but it is still useful to consider it as one *possible* perspective on ethnomethodology, which supports the idea that ethnomethodology can generate enquiries into the conflictual rationalities of government. Furthermore, the advantage of McHoul's Derridean perspective is that it is particularly sensitive to disambiguity and, by implication, to resistance.

The starting point of McHoul's argument is that *Of Grammatology* is an important piece of *critical* theory. Accordingly, whereas he states that the concept of 'critique' is *tendentious* in a Garfinkelian sense (that is, the concept cannot and should not be stipulated in advance, but is rather accomplished on an ongoing basis), he points out that his approach to critique is broader than the one patented by the Frankfurt School, namely an approach where 'critique' is supposed to invoke the irreducible *différance* of any linguistic expression (105). Whereas

5 Only a few researchers have attempted to draw on ethnomethodology with an explicit hope to be critical. Freund and Abrams (1976) represent a notable exception.

Derrida's notion of *différance* is supposed to critique the persistent privileging of speech over writing ('logocentrism')[6] and to play on the fine distinction between the sensible and the intelligible, McHoul draws on the term because of its intended ambiguity, playing on the double meaning of the French 'différer' which means both 'to defer' and 'to differ'. Hence, the critical potential of the term is its post-structural double insistence that signs have meaning only via their *difference* from other signs and that relational work is always required to determine the in situ *meaningful* difference. Meaning, then, is an intertwined phenomenon of observable difference and eternal deferment (105).

Subsequently, McHoul's approach to critique is derived from *crisis* rather than from a conception of criticism as social and/or moral judgement. McHoul understands critique as referring to a form of analysis that insists on meaning as a twofold phenomenon. Hence, whereas the meaning of any linguistic sign is *in principle* essentially unstable, in *practice* this instability nevertheless has to come to a turning point. Furthermore, since 'no chain of signification' can continue infinitely in practice, at some point a *crisis* will occur, and this point is the object of critical analysis (106).[7]

According to McHoul, such an understanding of critique can be related to ethnomethodology through the notion of indexicality. He understands the notion of indexicality to imply the same critical potential as the concept of *différance*. Taking inspiration from Derrida's notion of an *indefinite* referral of sign to sign, McHoul challenges the fact that the typical approach to indexicality focuses on the practical *definiteness* of meaning. In other words, he challenges the ethnomethodological tendency to focus on meaning-making as an unproblematised process in practice. Consequently, his argument for a critical ethnomethodology is founded in a partial rethinking of the concept of indexicality "from 'within' EM and 'between' EM and critical theory (CT)" (109). As indicated, this rethinking is motivated by the observation that ethnomethodology and critical theory share the notion of an inherent instability or indexicality of meaning. Hence, McHoul fully acknowledges Garfinkel's solution to the problem of social order as the entirely unique observation that "every stable social meaning (homogeneity) is always the effect of muddling through, making do, with unstable discursive conditions (heterogeneity)" (110). However, McHoul does not consider this approach that different from the approach to meaning-instability

6 In French, différance is pronounced *difference*; that is, it is pronounced identically with the correct French form *difference*.

7 'Chain of signification' is, of course, once again a reference to Derrida's post-structural elaboration on Saussure's structuralism.

articulated through the notion of *différance*. On the one hand, ethnomethodology attempts to demonstrate how indeterminacy is *worked* or *overcome* in practice and how it is repaired for the moment, "thereby generating positive social meanings or 'practical understandings'" (110). On the other hand, critical theory focuses on instability as an inherently discoverable phenomenon to be pointed out or even celebrated and utilises it "to explore and exploit the reflexive relation between civil debate and action, on the one hand, and theoretical debate and action on the other" (110).

Following McHoul, then, the difference between ethnomethodology and critical theory is not principal but related to how they deal with the same observation of indexicality/instability. In fact, his point is that ethnomethodology could prosper from considering how instability is dealt with in critical theory. In more detail, he suggests that notwithstanding all attempts to the contrary, ethnomethodology seems to relapse into a priori presupposing. He develops this point through a direct critique of Heritage's (1984b: 144) celebrated formulation that the ethnomethodological approach to indexicality is a matter of making definite sense with indefinite resources. In opposition, McHoul says: "It is not impossible to conceive of (or indeed to discover) practical situations where quite *in*definite senses are permissible or even required" (McHoul, 1994: 111). To exemplify, McHoul mentions situations ranging from popular texts such as the TV programme *Twin Peaks* to bold hypothesising à la Popper.[8] Against this, he observes that while indexicality has often been conceptualised in terms of meaning gaps (112), he would rather conceptualise it as a hypothetical point (113). In particular, he suggests that the indexical potential would be indefinite at the hypothetical point and that it would be possible to imagine at least two lines emerging from it. One line would carry the expression towards a point of ultimate closure where it could mean only one thing, and another line would carry the expression towards ultimate dispersal where it could mean anything. Whereas McHoul points out that an expression never in practice follows through any of these lines, he wants to make a case for at least a theoretical awareness of such lines. He remarks that it should be observed that the practical limits as well as members' related questions are quite different along each line. Along the line towards definite sense, limits and questions are concerned

8 Like McHoul, other researchers such as Wittgenstein and Pollner have pointed to similar insights. For instance, Wittgenstein has suggested that sometimes the indistinct picture is exactly the appropriate situation for all practical purposes, and Pollner has suggested that in some situations 'reality disjunctures' are treated as part of the routine business of life as usual (Wittgenstein, 1998: § 71; Pollner, 1987: 69).

with when enough is enough for all practical purposes (114). Along the line towards ultimate dispersal, limits and questions are concerned with how to invent and speak more differently (114).

Accordingly, McHoul suggests that a critical ethnomethodological study would be a study that eschews the a priori assumption of definite sense and investigates the overall horizon of possibility. Such a study would be open to the possibility that some practices might have "such things as disorder, conflict, contradiction, struggle, antagonism (and so on) not merely as actional achievements but *as part of their taken-for-granted background and foundation*" (114, emphasis in original). Hence, in opposition to most ethnomethodology, McHoul suggests that ethnomethodology should acknowledge that contingently accomplished interaction is always afforded by and affording certain politics. Not only does this come off as a clear invitation to connect ethnomethodology with studies of governmentality, but, considering McHoul's emphasis that disorder, conflict, contradiction, struggle and antagonism are constitutional of interaction, it more specifically comes off as an invitation to connect ethnomethodology with an understanding of resistance as a constitutive moment of government in studies of governmentality. That is, McHoul suggests that contested categories are understood as matters of 'political' struggle.

In summary, then, this chapter has introduced a research domain that has a *theoretical* focus on rationality and morality similar to the one in studies of governmentality, and, as an *inherent* part of this theoretical argument, it has suggested some methodological guidelines for the investigation of how such rationality and morality are accomplished. That is while the chapter has been framed as theoretical, it has indicated a framework for studying governmentality at the molecular level. Or, in a wording reminiscent of studies of governmentality, the chapter has indicated how the ongoing accomplishment and resistance of mentalities of government can be made available for enquiry at the molecular level.

4. Considering commensurability

It is time to explicate what I have so far presupposed more implicitly, namely that it is both theoretically sound and fruitful to connect studies of governmentality and ethnomethodology. Thus, this chapter is not merely the last thread of the three-stringed strand that makes up the theoretical argument of the book; it is the thread that illuminates how the threads can in fact be seen as one interdependent strand.

In the first thread I introduced and discussed studies of governmentality, focusing on their sensitivity to current forms of governing, especially contemporary governmentalities in which power and freedom cooperate through the notion of the conduct of conduct. However, I also pointed out that such studies may have an inherent tendency to end up presupposing and homogenising its findings, treating current liberal governmentality as a naturalised end point rather than as a contingent and unstable, continually contested accomplishment. I observed that even though it is often pointed out that an inevitable aspect of carrying out studies of governmentality is an investigation of the *hows* at the molecular level, recommendations of how to do so are few and studies that actually follow the ambition through are to my knowledge non-existing. Lastly, I suggested that the lack of an observable attention to the subtleties at the molecular level might be related to the failure to acknowledge that governing is inevitably co-constituted by resistance.

Considering these insufficiencies of studies of governmentality, the second thread was motivated by the ambition to connect studies of governmentality with ethnomethodology; that is, with a theory that both acknowledges the importance of studies of the *hows* at the molecular level *and* is co-constituted by a method that demonstrates how this can be done. Hence, in the second thread I introduced and discussed ethnomethodology. However, whereas ethnomethodological research is unsurpassable when it comes to 'tools' for studying such *hows*, not all such research is equally interested in how rationality and morality are co-constituted in category work, and some openly reject the 'politicisation' of ethnomethodology implied by an approach that holds that there is no exit from the moral order, neither for members nor for analysts. Consequently, I outlined not just any version of ethnomethodological research, but a line which can be connected with the politicised ambition within studies of governmentality to approach the ongoing negotiations of rationality and morality as part and parcel of government. This specific line of ethnomethodology carries on and develops the spirit of Garfinkel's original work and Sacks's early work, maintaining both the interdependent relationship between theory and methodology and the interdependent relationship between category and sequence. Hence, I have discussed a theory that can be connected both with my general interest in automobility as a problem of governance and with my more specific interest in governing as being co-constituted in talk-in-interaction through the accomplishment of rationality and morality at the intersection of a municipal strategy for 'improving' citizens' transportation practices and citizens' self-administration. I have also discussed a theory which grounds and invokes a method for studying this discursive accomplishment.

However, while I have indicated how the two theories can be connected, it remains to be discussed more systematically that such a connection is *theoretically* sustained. Accordingly, it is time to take a step backwards and consider the commensurability of the theoretical foundations of, at one end EMCA and, at the other, studies of governmentality. In other words, the purpose of this chapter is to discuss the feasibility of connecting Garfinkel's ethnomethodology and Foucault, or, in Laurier and Philo's (2004) words, to consider the arguments for "the peculiar appropriateness of reconciling ethnomethodology with the work of Foucault" (422).[1] That is, the chapter discusses the arguments of the most detailed accounts provided so far, concerning the theoretical feasibility of connecting ethnomethodology and Foucauldian studies.[2]

Whereas only a few researchers have considered the theoretical feasibility and productivity of juxtaposing Foucault and Garfinkel, McHoul (1986; 1996) and Laurier and Philo (2004) have done thorough theoretical work on ethnogenealogy and ethnoarchaeology, respectively.[3] On an overall level, the researchers themselves treat their choice of focus on genealogy and archaeology as being of

---

1 Whereas Laurier and Philo talk about *reconciliation*, I prefer to talk about *connecting*, considering the fact that ethnomethodology and Foucauldian studies have never really been related before.

2 However, besides the more thorough accounts discussed in this chapter, others have indicated the possibility of connecting EMCA and studies of governmentality in particular (see, for instance, McIlvenny, 2008 and 2009; Klausen, 2012) and ethnomethodology and Foucauldian studies in general (see, for instance, Miller, 2004; Tate, 2007).

3 However, Lynch (1993) should also be mentioned. Whereas Lynch does not juxtapose Foucault and Garfinkel as such, he does briefly consider the prosperous opportunities of connecting the two, and he should definitely be acknowledged as one of a few figures in ethnomethodology who are willing to hybridise it with other approaches, if cautiously and critically (131). It should be noted, however, that to my knowledge nobody else has done any detailed work with a view to connect these two traditions.

no particular relevance. Hence, McHoul tends to use archaeology and genealogy interchangeably,[4] and Laurier and Philo remark that they consider the methods of genealogy and archaeology closely related. Furthermore, Laurier and Philo point out that they *sympathise* with McHoul and make no 'special point' in calling their follow-up to his lead 'ethno*archaeology*' (422–3). Consequently, when I discuss McHoul and Laurier and Philo in the following, I do so without considering whether they call their respective works ethnogenealogy or ethnoarchaeology.

McHoul's (1996) interest in connecting up Foucault and ethnomethodology is motivated by his ambition to assemble an "effective semiotics"; that is, a semiotics that is "able to range from the most basic of semiotic practices (the sheer intelligibility of signs or methodic activities), via the level of socio-logical problems and solutions, to the historical embeddedness (and supra-situational availability) of semiotic techniques within communities" (101). Accordingly, McHoul observes that ethnomethodology and Foucauldian archaeology/genealogy represent the "two extremes of social theory" which such a semiotics must encompass and, by implication, reconcile (101). Or, in other crudely approximating words, McHoul (1986) points out that whereas a Foucauldian approach can afford a more dispersed *diachronic* perspective, ethnomethodology affords a uniquely unimposing *synchronic* analysis of order as it is continually and reflexively accomplished (66). Similarly, Laurier and Philo (2004) point out: "Foucault provides a historicism which ethnomethodology lacks. From the other side, ethnomethodology localises what otherwise risk becoming grand periodisations within Foucault's histories" (423).

In line with this, McHoul and Laurier and Philo present similar reasons for why it might be fruitful to connect the two different perspectives. On the one hand, Foucault and Garfinkel each represent their own superb approach. On the other hand, both 'extremes' face some lurking problems. Whereas the broader, predominantly diachronic Foucauldian studies are always at risk of overlooking discontinuities and contingency, ending up homogenising and reifying, the narrower, predominantly synchronic ethnomethodological studies cannot account for power struggles that are dispersed into domains that are stretched beyond the in situ observable. Consequently, it is exactly because of Foucault's and Garfinkel's radical differences that McHoul as well as Laurier and Philo find it not only interesting but also important to connect the two.

I generally agree that it is motivating to connect an ethnomethodological and a Foucauldian perspective due to the different strengths of the two theories in 'synchronic' and 'diachronic' analyses. However, considering for a moment the

4 See, for instance, McHoul, 1996: 101.

particularities of the versions of ethnomethodological and Foucauldian research that I draw on, I would like to make two more specific points. Firstly, the ethnomethodological work on interactional category work should be considered. Whereas ethnomethodology focuses primarily on conducting locally situated analysis, it should be recalled that the notion that categories are accomplished for another first time *also* implies that categories have certain histories. Recalling Jayyusi, I would like to point out that her interpretation of ethnomethodological indifference recognises that synchronic and diachronic analyses necessarily co-constitute each other. Hence, when Jayyusi argues that disinterested description can never reach beyond the 'moment' of description and, by implication, that the analyst has to draw on the resources of his or her own members both in order to locate the moment and to connect it with the observations made in other disinterested moments, she can be said to argue that in order to make sense, all 'synchronic' descriptions depend on 'diachronic' embedding. Similarly, I do not draw on any version of Foucauldian studies, but on current studies of governmentality that pay explicit attention to the mentalities of government as they are accomplished here and now at the molecular level (cf. Rose, 1999).[5] Hence, my starting point is not that EMCA and studies of governmentality represent 'each their extreme' position, but merely that the one focuses primarily on the situated accomplishment whereas the other focuses primarily on the contingent and discontinuous 'development' through longer time spans. I draw on the more diachronic insights of studies of governmentality in my approach to automobility as a problem of governance co-constituted by the accomplishment of rationality and morality, and I draw on the more synchronic expertise of ethnomethodology to attain knowledge of how automobility-maintaining governmentalities are accomplished here and now. I do not understand these two perspectives as being different in kind, only in focus, but I do agree with McHoul and Laurier and Philo on the general point that it is this different foci of the two that make it particularly interesting to connect them.[6]

---

5 See also Biebricher (2008). Discussing that future studies of governmentality should focus more on struggles, discontinuity and contingency, Biebricher suggests that this could be achieved if such studies develop "a more heterogeneous account of contemporary governmentalit(ies), both synchronic and diachronic, [addressing] the teleological as well as the presentist aspects of Foucault's account" (396).

6 Hence, in parallel to the Latourian dissolution of micro and macro, I do not conceive of the diachronic and the synchronic in terms of the linear versus the punctual, but in terms of stretchable moments within one and the same ontological assemblage.

Of course, both McHoul (1986) and Laurier and Philo (2004) are well aware that an argument for connecting Foucault and Garfinkel is not sufficiently sustained by simply outlining a useful difference in orientation. Hence, they further point to obvious commonalities between Foucault and Garfinkel, thereby emphasising that despite the fact that the two have only rarely been connected, they do in fact have a rich common ground for commensurability. Firstly, McHoul observes a 'problem/solution metaphor' which can be a point of departure for explicating 'family resemblances' between Foucault and Garfinkel (65). More specifically, McHoul discusses to what extent Garfinkel and Foucault share an understanding of problems as something which members necessarily deal with through social action or conduct. McHoul remarks that such an approach is observable in ethnomethodology, for instance, when Sacks treats a phenomenon like conversational sequencing as a *problem* which members have to face by invoking collectively available solutions. Furthermore, McHoul discusses how the metaphor can be applied to Foucault's work. This is less straightforward, but McHoul observes that Foucault does utilise a concept of "Discourse as problematics" as well as a concept of "discourse as solution" (66). In greater detail, McHoul remarks that whereas Foucault's earlier work displays little interest in solutions and techniques for accomplishing everyday life, his later work does. In his earlier work, Foucault is predominantly preoccupied with the domain of the orders of discourse per se. Here he approaches the fields of madness, civilisation or knowledge merely as problematics or as contingent, but somehow solidified discursive sediments. On the other hand, in his later works, he is profoundly interested in how the subjects sedimented by the discourses must accomplish their problems using the very same discursive resources. McHoul mentions Foucault's work on the stories of the murderer Pierre Rivière and the hermaphrodite Herculine Barbin as concrete examples of this interest (66). Against this, McHoul concludes that Foucault can be said to be sharing an ethnomethodological interest in how everyday life problems are accomplished, solidifying discursive fields for another first time.[7]

---

7 McHoul puts this last point slightly differently and slightly closer to Foucault's and Garfinkel's respective wordings: "Foucault talks here of 'eventualisation' while Garfinkel talks of invariance (Discourse) as a function of contingency (discourse)" (McHoul, 1986: 66). Note, however, that McHoul also emphasises their similarities, as he in brackets points out how the Discourse/discourse distinction is applicable also to the work of Garfinkel. Further, while McHoul does not mention Foucault's lectures on governmentality in this argument, I would like to add that Foucault's attention to the population as being not merely the subject of needs but also the object utilised for fulfilling these needs in a similar way points to Foucault's heightened interest in the

Furthermore, this common attention to reflexively accomplished techniques of everyday life points to yet another significant point for connecting Foucault and Garfinkel: their identical ways of rejecting notions of reifying social structures. Accordingly, McHoul observes that the shared rejection indicates that Foucault's treatment of truth can be connected with Garfinkel's notion of the socio-factual (66). Following McHoul, then, it is theoretically sound to approach the Foucauldian notion of truth through ethnomethodologically informed analyses of the reflexively accomplished social order. Neither Foucault nor Garfinkel searches for truth or rationality behind the surface of actual observable interaction, and they both approach the domain of the true or the socio-factual as the domain of the contingently accomplished at a given time and place. Also Laurier and Philo (2004) make this observation when they point out that neither Garfinkel nor Foucault is predominantly interested in what might be behind the curtain, but they are both interested in all the little details of the surface. In their pointed wording: "There is, if you will pardon the mild paradox, a deep concern with the surfaces of the world" (429).

McHoul pursues the argument, pointing out that with regard to language, Foucault and Garfinkel equally reject aprioristic theorising on unobservable and *deep* structures. As McHoul (1986) observes, even though Garfinkel's discourse on discourse is infused with Anglo-American empiricist notions of language use and speech (in Saussurean terms *parole*), whereas Foucault's bears resemblance to continental concerns with language as such (*langue* with Saussure), both of them eschew any kind of structural starting point in *la langue* and insist on the importance of paying close attention to actually accomplished *parole* (67). Relatedly, I would like to add that neither Foucault's nor Garfinkel's approach to language is comparable with the contemporary development in speech act theory. Whereas all three directions share the fundamental assumption that language is not merely a matter of representation, but rather of social action, Foucault and Garfinkel again insist on the importance of paying close attention to actually accomplished (verbal) interaction. In this way, Foucault and Garfinkel circumvent the speech act theoretical approach which starts with this or that class of action and then analyses it and the conceptual components that make some particular act an instance of that class.[8]

continuous discursive negotiations and struggles which are an inevitable part of accomplishing, maintaining and managing the problems of everyday life.

8 I have elsewhere discussed the difference between Foucault's *statement* and Searle's *speech act* (Lindegaard, 2010).

Closely related, Laurier and Philo point out that Foucault and Garfinkel can be said to take similar approaches to 'local rules' as opposed to 'transcendental laws'. They note that this common ground is too often neglected because of a too simplistic understanding of Foucault's rejection of formalism and ethnomethodology's *apparent* focus on it (424–7). Laurier and Philo agree that it implies a rejection of formalistic analysis when Foucault insists that "what is said *is* what is said" (425–6), but they emphasise that even though ethnomethodology is concerned with formal structures of practical actions, it has to be acknowledged that it is a very odd or peculiar interest in formal properties that is thereby invoked (426). Accordingly, they suggest that any 'formalism' in ethnomethodology boils down to an interest in rules as always 'used for another first time', that is, as continually and reflexively accomplished *for all practical purposes*. Against this suggestion, they emphasise that *ethnomethodological* formalism might not be too distinct from Foucault (426–7). Furthermore, Laurier and Philo point out that not only do both Foucault and Garfinkel reject a reifying formalism and all notions of transcendental laws, they also have a similar interest in rules as a "specific, local, and mutable condition of *existence*" (427). Laurier and Philo remark that this notion of rules is developed most extensively within ethnomethodology, but they add that Foucault touches upon it, for instance, when he discusses how statements are somehow rule-governed and dispersed into *discursive formations* (Foucault, 1972: 41–2). Hence, Foucault can be said to be embracing an understanding of rules and laws similar to that of ethnomethodology. Following this, Laurier and Philo stress that it is *not* the job of the analyst to *do* the defining or classifying (427). Rather, the analyst should be content with pointing them out. Or, in ethnomethodological wording, categories should be a member's, not an analyst's, concern.

5. Site, data and approach

This chapter is the first of three discussing and demonstrating how the theoretical argument of the book can be utilised in actual, empirical analysis. Considering the empirical site, the data and the approach undertaken, this chapter provides the methodological clarifications needed for the analysis proper carried out in Chapter Six and Chapter Seven. Particularly, this chapter clarifies how I approach the discursive accomplishment of rationalities that co-constitute the governing of automobility; or, specifically, it clarifies how I approach the rationalisations and negotiations of, firstly, one's own transportation practices and, secondly, the intersection of a municipal strategy aimed at 'improving' citizens' transportation practices and citizens' self-administration.

Whereas the methodology is mainly informed by the specific approach to EMCA discussed in Chapter Three, naturally it reflects the dual theoretical starting point in ethnomethodology *and* studies of governmentality. Accordingly, the methodology does not simply reflect a 'pure' EMCA study of current forms of government, but an EMCA-inspired analysis utilised to make a first stage analysis of orders of rationalities that is then interpreted in terms of governmentality theory. As such, the analysis is clearly designed and carried out by an analyst who is reflexively embedded in theoretical as well as empirical memberships. The chapter comprises three sections; the first presenting both the empirical site and the design of the study, the second undertaking a discussion of how to approach focus group data in ethnomethodological research, and the third presenting the most important tools utilised in the analysis.

Land of Opportunity: Energy Village Horslunde

Horslunde is a village with 750 inhabitants in a decentralised, demographically vulnerable part of Denmark, namely the island of Lolland. The village is the main village in its area, and until the latest local government reform in Denmark (2007) it was the capital town of Ravnsborg Municipality. Since the reform, all municipal activities and offices have been transferred to the new capital town of the new, much larger Lolland Municipality. However, Horslunde is still a village in which most everyday needs can be fulfilled; for instance, it has a supermarket, a local shop and a bakery, two schools (one private, one public), a kindergarten and a care home for the elderly. Furthermore, it has craftsman shops, an art society, leisure and sports associations as well as a church. However, there are few jobs in the village, and most adults commute to work in other places, as do teenagers attending upper secondary school.

The infrastructure in the area of the village is inherently automobility-generated and automobility-generating. From a Danish perspective, the village must be characterised as rural and as placed in a highly decentralised municipality. Although it is set only ten kilometres from the closest larger town, Nakskov, it is pointedly a *town*, not a city. For instance, there are almost no education opportunities beyond upper secondary school level, and even though public transit to other parts of the country can be initiated in Nakskov, it is not possible to commute directly to any larger city. Furthermore, busses between Nakskov and Horslunde are rare and are not operated after nine pm on weekdays, and they are operated only a couple of times at weekends. Accordingly, life in Horslunde is lived in a car-dependent way.

Horslunde is indeed a site for a study of current forms of governing. The village is designated a 'demonstration project' in the project *Land of Opportunity* which is a six-year partnership project between the strategic, philanthropic foundation Realdania and three Danish municipalities, one of which is Lolland Municipality. The partnership project is inspired by the OECD's construction of *The New Rural Paradigm* (Crosta et al., 2006) which is the result of the observation that a traditional focus on financial redistribution and agriculture-based policies is "not able to harness the potential" of a new globalisation-generated "range of economic engines for growth" (3). Following this observation, the OECD recommends a new approach to rural policy, requiring a new set of challenges: "It requires a less "defensive" approach to rural policy and stronger co-ordination across sectors, across levels of government, and between public and private actors. It also requires a new focus on places rather than sectors and an emphasis on investments rather than subsidies" (3). Significantly, the new paradigm focuses on government as a dispersed process at all levels and, again, on *investments*, not subsidies. Hence, the new approach to development seems to focus on how the *freedom* of the governed can be utilised as a resource in generating economic growth.

Project *Land of Opportunity* is formulated as part of this new development philosophy, in that it represents a project-oriented approach to development in rural districts and an intense focus on local anchoring amongst citizens and stakeholders (*Land of Opportunity*, project summary, 2009: 3). The project is not "primarily concerned with the creation of economic growth", but rather aims at maintaining and developing the qualities and potentials already to be found in the particularities of individual places (3). Hence, the starting point is that development projects in rural areas need to take into account the fact that such areas are far from identical. It is on this basis that the three partnership municipalities have initiated "demonstration projects which try out new strategies for creating development and a quality of life in rural areas" (3).

In brief, project *Land of Opportunity* was initiated in 2006 with a number of initiating research studies and analyses. In 2007–8 the project continued with a project development phase in which demonstration projects were developed and approved. The main phase, the realisation phase, has run since 2009 and was to be concluded in 2012. The total budget for the project is DKK 195 million. Realdania has contributed with DKK 105 million and each of the municipalities has contributed with up to DKK 30 million.

Focusing on Lolland, project *Energy for local development* is but one of four selected projects. This selection is the result of a relatively top-down-generated dialogue between the citizens and the municipality (På vej i Mulighedernes Land, 2008: 42). More specifically, the municipality made a call for ideas for projects that would fit with at least one of seven proposed development categories. One of these categories was related to the municipality's general energy strategy. In slightly more detail, Lolland Municipality is a so-called 'climate municipality'. Firstly, in 2007 the municipality signed an agreement with the Danish Society for Nature Conservation to reduce the municipality's CO2 emissions by three per cent annually until 2025. Secondly, in 2009 the municipality signed a second agreement, this time with the Sustainable Energy Action, which is an initiative under the Covenant of Mayors.[1] Relatedly, the municipality has devised a climate and energy strategy, pointing out the strategic ambitions within the three focus areas: (1) climate change mitigation through CO2 reductions, (2) adaptation to climate change and (3) green growth initiatives. Accordingly, when the citizens of Horslunde proposed that their village could be transformed into an *energy village*, the municipality chose this project because it is readily relatable to the focus on climate change mitigation through CO2 reductions; in fact, the project *Energy for local development* provided the opportunity to further the municipality's ambitious energy visions via a focus on citizens' everyday energy practices and, "at the same time, [it] provide[d] an image boost for Lolland" (Project Summary, 2009: 7).

---

1 The Covenant of Mayors is the mainstream EU movement involving local and regional authorities voluntarily committing to increase energy efficiency and the use of renewable energy sources in their territories. By their commitment, Covenant signatories aim to meet and exceed the EU twenty per cent CO2 reduction objective by 2020.

A core group of citizens formulated the original project idea, but the municipality has been in charge of both the concrete design and the execution of the project (the design and the concrete initiatives have required ongoing approval from Realdania). The realisation phase was initiated in early 2010, marked by a call for citizens to fill in a questionnaire concerning their domestic energy consumption. The questionnaire responses were used for a general mapping of energy use in the village, and the questionnaires can be seen as related to one of the core dimensions of the project, namely the establishment of a pool of money which the citizens can apply for if they want to make energy-efficient renovations of their houses. Additionally, the energy village project has featured a visit from 'the energy service'[2] which resulted in different renovation opportunities being pointed out at the two schools and in a private home, and in October 2010 there was a large energy exhibition in the hall of the public school, similarly focusing on the different options in relation to energy renovations and developments in private homes. Furthermore, October 2010 featured the first of five so-called 'theme months' focusing on different opportunities for energy savings in everyday life. Hence, October 2010 featured a focus on domestic heating, November 2010 focused on transportation practices, December 2010 focused on daily waste management, then, after a winter break, March 2011 featured an electricity theme and, lastly, in April 2011 the theme was water usage. The transportation theme month, specifically, attempted to inform citizens about different 'greener' and more energy efficient forms of daily mobility; that is, it attempted to conduct the transportation conduct of the citizens of this car-dependent village. This was realised through a four-page transportation flyer informing them of different aspects of greener driving and through a transportation theme day at which it was possible to talk to an expert on green driving and to participate in a contest with the opportunity of winning a course on green driving. The flyer was displayed and made available at different spots in the village, such as the baker's shop, and the theme day was carried out on a Saturday in the parking lot in front of the supermarket.

Following my two-fold understanding, firstly, that automobility is predominantly a problem of governance, particularly in the face of climate change, and, secondly, that this problem is co-constituted by the ongoing accomplishment of

2 The energy service (Energitjenesten) is an initiative under Sustainable Energy (Vedvarende Energi) which is a public organisation that has been aiming for 100% renewable energy in Denmark since 1975. The energy service is organised around ten local departments in the country, focusing on and informing about energy savings and different forms of sustainable energy sources and solutions.

rationalities, Energy Village Horslunde seemed an obvious case for a study of how rationalities that co-constitute automobility are accomplished at the intersection of a municipal strategy and citizens' accounts for their own conduct. Not only is life in Horslunde conducted in a way that depends heavily on cars as a mode of transport, the municipal strategy is embedded in project *Land of Opportunity* which exhibits an explicit ambition to govern through the conduct of conduct of the governed. In other words, it exhibits a way of governing that is reminiscent of the ways pointed out by recent studies in governmentality. Considering my more specific decision to focus not just on any form of talk-in-interaction, but on focus group interaction, it is a consequence of Rose's observation that focus groups can themselves be seen as a tool of governmentality. In more detail, Rose (1999) observes that focus groups can be seen as one of a whole array of devices and techniques that have been invented to make current communities real (189), and he points out that in focus groups, citizens are called upon and comply as they map out new spaces of culture, clarify current values and virtues and insert them into the deliberations of authorities (189). Accordingly, I understand focus groups as an accounting-oriented device that magnifies the ways and doings of rationalities; that is, as a vehicle that emphasises the call for pointing out and accounting for positions.

Against this, the outline of my focus group study should briefly be considered. The study comprises two sets of sessions with the same groups of citizens and with me as moderator.[3] The first set was carried out before the municipal transportation theme month, and the second set was carried out after this month. In more detail, in the sessions constituting the first set, I asked broadly about the participants' everyday transportation practices. Hence, I quite simply asked about the four main forms of transportation undertaken in the village, namely walking, bicycling, driving (a car) and using public transportation. On the other hand, in the sessions constituting the second set, I asked more specifically about the citizens' reactions to the municipal transportation theme month, focusing on how the citizens related the strategy to their own practices. Hence, at the most general level, I asked about the participants' overall experience of the transportation theme month, about their experience of the transportation theme day and about their views on green driving as suggested by the municipal flyer. The ambition of this design was that both sets of sessions would make rationalities and moralities that co-constitute the governing of automobility available for enquiry, but that they would differ; the first set would more prominently

3 I return to my participation in the study as moderator in the next section.

demonstrate how rationalities are accomplished in the face of a simpler call to account for one's own practices, whereas the second set would more prominently demonstrate how rationalities are accomplished in the face of a more complex call to account for one's own practices in relation to a strategy aimed at changing those practices. In other words, it was my ambition that the first set would more prominently display the *hows* of the practices concerned to conduct one's own conduct, and that the second set would more prominently display the continually contested intersection of the practices concerned to conduct the conduct of others and the practices concerned to conduct one's own conduct. Accordingly, my commonsensical starting point was that the second set of sessions would most prominently feature the *hows* of the contradictory role of resistance and negotiation as both the boundary and the constitutive moment of governing.

Focus group data and ethnomethodology

Scholars of ethnomethodologically informed discourse studies are often sceptical of the use of interview data such as focus group data. Some scholars quite simply reject interview data with reference to a general preference for so-called naturally occurring data. Other scholars acknowledge that interview data can be of some use if the distinction between natural and contrived data is given up and replaced with a distinction between interview data as topic or as resource. In greater detail, such scholars argue that interview data is perfectly adequate data if the researcher wants to study the topic of interview interaction, but inadequate as data for studying phenomena that go beyond the phenomenon of interview interaction. Neither of these more and less sceptical positions are, at face level, surprising, considering the ethnomethodological commitment to study social order as accomplished in situ, not as something that pre-exists or goes beyond the situated interaction. This book, however, challenges not only the first, but also the second position and suggests that it is, after all, possible to do committedly ethnomethodological studies of focus group data that demonstrate how members to a focus group setting accomplish governmental rationalities, and, significantly, how they do so by utilising certain available resources.[4]

Whereas some scholars within ethnomethodologically inspired research, such as discursive psychology and conversation analysis, seem to maintain a distinction between so-called natural and non-natural data (preferring the former

4 This argument, which is extended in the following, is also unfolded in Lindegaard (2014).

to the latter),[5] other scholars rather concern themselves with the distinction between resource and topic. For instance, Rapley (2001) suggests that interviews should be seen as sites for studying how members produce "talk-about-a-topic-*in-interviews*" (318, emphasis in original), suggesting that his own interviews with both drug and non-drug users should be seen as sites for studying "how-people-talk-about-drugs-*in-facilitative-and-neutralisitic-interviews*" (318, emphasis in original). Along the same lines, Houtkoop-Steenstra (2000) has shown how survey talk is accomplished; Watson and Weinberg (1982), Heritage and Greatbatch (1991), Widdicombe and Wooffitt (1995) and Mazeland and Have (1996) have shown how interview talk is accomplished; and Puchta and Potter (1999; 2004) have shown how focus group talk is accomplished. Furthermore, Hester and Francis (1994) and Baker (2004) have argued that interview data should be approached from a perspective that draws not only on conversation analysis, but also on membership categorisation analysis, emphasising that interviewing should be understood as an interactional event in which members draw on their cultural knowledge, including their knowledge about how members of categories routinely speak. Hence, Baker (2004) observes that questions should be seen as a central part of the data, shaping how and of which categories the respondents should speak, and that interview responses should be understood as the work of a member of a category accounting for activities linked to that category (163).

Whereas all these studies transcend the discussion of *naturalness*, the orientation to interview data as *topic* often implicates the seemingly obvious assumption that interview data is treated as adequate for obtaining knowledge about interview interaction but inadequate for obtaining knowledge of social phenomena that have any relevance *beyond* the interview setting. However, considering the argument that yes, social order is contingently achieved for all practical purposes, but, importantly, this is *not* the same as saying that certain members in a certain setting could ever accomplish just *any* order, it is clear that also analyses of interview interaction have the potential of pointing out phenomena that go beyond the narrower focus of the structure and membershipping constituting interview interaction; for instance pointing out certain *readily available* resources in the ongoing accomplishment of order. A few other scholars have indicated similar, broader understandings of which insights we can hope to get

---

5 See, for instance, Have (1999: 48ff), Heritage (1984b: 234ff; 1988), Heritage and Atkinson (1984: 2–5), Potter and Wetherell (1995) and Silverman (2001: 159ff) for the maintenance of such a distinction.

from interview data. Rapley (2001), for instance, touches upon the significance of the 'availability' of certain discourses, identities, narratives, repertoires, rhetorics etc. in interview settings (318), and when Grancea (2010) speaks of participants' social experiences (165), he, too, seems to assume that his interview data informs us about something more general than the structure of interview interaction.[6] However, the question has never been debated in much depth, and, consequently, neither have there been any major, committedly *ethnomethodological* studies of focus group interaction that are concerned with, for instance, how the in situ accomplishment of rational order could tell us something about dominating rationalities in particular societies.

Additionally, I should like to recapitulate one more conventionally troubling question concerning interview data: the question concerning the feasibility of researchers who are shifting between doing being moderators and doing being analysts.[7] Within ethnomethodological research as such, there is an observable preference for data that the researcher has not been co-producing. The resistance to researcher involvement is perhaps strongest within discursive psychology (cf. Potter, 1997), but within conversation analysis, Have (2002) suggests that researchers should uphold a distinction between 'natural' and 'contrived', methodologically speaking. In contrast, this book follows Speer (2002b) as she emphasises that it is, of course, not possible to separate ontology (or epistemology) from one's methodology, (546) and that the resistance to researcher involvement entails a not very ethnomethodological reification of the researcher (Speer, 2002a: 518). Apparently, no one has discussed the question of researcher involvement in any detail since Speer,[8] but Whitehead (2012) briefly touches upon the question, countering what he presents as Stokoe's (2012) claim that membership categorisation analysis research should follow the insights of conversation analysis and concern itself with naturally occurring data that has not been generated by the researcher (339). Accordingly, I would like to suggest that the question of researcher involvement is still in need of discussion and, specifically, of demonstrations of how researcher involvement can be approached in actual analysis.

Following Speer, then, this book takes it as its starting point that it does not represent a necessary problem for researchers to use data they have themselves

---

6 Referring to Grancea's study, also Whitehead (2012) seem to be open to the possibility that interview data can provide knowledge of phenomena reaching beyond the activity of interview interaction.

7 Of course, the questions of naturalness and researcher involvement are closely interrelated. Also Speer (2002a) remarks this.

8 See, however, Griffin (2007) for a single, notable contribution to the debate since 2002.

generated – such as data generated by focus groups, moderated by the researcher her- or himself. Yet, I would like to suggest that it should be recognised that in such cases, 'doing being researcher' is likely to involve (at least) two different, observably relevant memberships: namely, 'doing being moderator' and 'doing being analyst'. Accordingly, as it will be made clear in the following analysis of focus group data, part and parcel of accomplishing 'doing being analyst' is to consider the interaction of the member 'doing being moderator' as interaction by one of the parties to the focus group setting.

Analysing focus group data

The remaining part of this chapter is concerned with a brief account for the most important tools utilised in the analysis. I will not mention every single feature of membership categorisation analysis and conversation analysis I draw on, but only those I have drawn on *throughout* the analysis, postponing discussions of less frequent aspects for when they become relevant to the analysis proper.

Concerning membership categorisation analysis, this approach to categories is quite distinct from a more common approach in which context is understood as an unspoken, reified parameter that informs or influences the interactional display of categories in conversation. In ethnomethodology, context is not imposed, it is achieved, and, by implication, membership categorisation analysis abstains from stipulating the existence or relevance of categories, unless members are evidently oriented to them. However, as discussed, the refusal to *ironise* or *judge* members' accounts is merely the negative side to a positive research programme that conceptualises categories as members' accomplishments. Hence, membership categorisation analysis affords moments of indifferent description by focusing on how *members* make themselves and others recognisable as specific kinds of people. Accordingly, the task of the analyst is to point out how 'recognisability' is observable, for instance, through the ways social categories are invoked through their different, co-constituting predicates.

Reiterating the features to accomplish this task, the most fundamental membership categorisation analytical tools stem from Sacks's basic observations on membership categorisation devices, categories, category-bound activities as well as the consistency rule and the economy rule. However, in addition to Sacks, I draw on the works of Sharrock, Jayyusi and Watson on the expansion of category boundness. Whereas Watson (1978) is one of the first to explicate that Sacks's notion of category-bound activities can be read as just one type of category-bound *predicates*, I particularly draw on Jayyusi's observations regarding the moral implications of all category work and Sharrock's observations regarding the rights, obligations and responsibilities invoked by owning knowledge (52). Hence, I talk primarily of category-bound *predicates*, not activities, and I understand the term as embracing any predicate that can expectedly be ascribed to the members of a given category, such as rights, entitlements, obligations, knowledge, attributes and competencies. Yet, it may still be difficult to comprehend the practical implications of such an approach, and perhaps the best way to demonstrate the complicated details is with an example from the analysis proper.

A crucial aspect in one of the analyses is that the moderator *accounts* for why she rides her bicycle when it rains when she does not enjoy it, invoking and subscribing to a category-bound 'no other alternatives but riding'. Briefly unpacking how this membership work is accomplished, it is necessary to start by considering what the moderator actually says. In exemplifying what she understands by 'always riding, also when it rains', she says that she always rides her bicycle to the train, and she adds that it does not matter whether or not it rains. She underlines that it is 'of course' no fun when it does rain, 'but' that she has no other alternatives. At this point, it should be noted that Sacks (1992a) observes that it is perhaps more often than not the case that social categories are made recognisable not through blatant naming, but through different forms of 'hinting' (578–9). Accordingly, to point out ongoing category work, one is 'allowed' to draw on one's members' knowledge to explicate what is going on in the data. In this particular piece of data, it becomes relevant that the moderator always rides her bicycle to the train and that she has no other alternatives. Most members would probably interpret this data to reveal that the moderator is a 'train commuter' and, perhaps, by extension, that she is 'an employee', 'a student', 'a client' or belongs to another category whose members are obligated to go somewhere on a regular basis. Furthermore, members would believe that train commuters have to get to the train station because they have to take the train, and that this particular commuter, the moderator, lives at a distance from the train station which makes riding a bicycle a possibility, and, importantly, the *only* possibility. That is, members would take it to mean that the moderator thinks that she does not have access to a car or public transportation for instance, or that she does not have the skills needed for utilising either of them. In short, we understand it as a self-identifying appropriation of a certain access and/or of certain skills.

Yet, even though this appears to be a plausible understanding, it still has to be made evident that it is actually the understanding that is made observably relevant *in data*. Here, it is helpful to recall Sacks's (1992a) observation that a common part of our everyday orientation and solution to problematic utterances is to assign a category to the person who performs the utterance; for example the category of 'people who are entitled to say such a thing' (180). Accordingly, it can be noted that the 'no alternatives' predicate is doing *justification*, since it is invoked to explain what the moderator *collaboratively* treats as a problematic activity, namely her production of the utterances 'I ride a bicycle when it rains' and 'I do not enjoy riding a bicycle when it rains'. In particular, it can be said that the purpose of the 'no alternatives' predicate is to establish that the moderator belongs to a category of persons who are entitled to say such a thing. It can also be said that the specific character of the category to which the moderator 'belongs', for example 'employee' or 'student', is not relevant for the interactional task at hand. Hence, it can be specified that the moderator subscribes to self-identification regarding certain access/skills, and that she thereby invokes a binding category incumbency, but it *cannot* and need not be specified which incumbency we are talking about. In more detail, while a membership categorisation analysis cannot specify whether the moderator belongs to any of the suggested categories (in fact, it is not even possible to categorise the moderator as 'moderator' or as a 'she' on the basis of the few details of the interaction reported here), it can point out that the moderator projects a moral order in which it is considered observably accountable to ride a bicycle when it rains, although it is not considered enjoyable. In other words, the analysis can demonstrate how culture in action is accomplished.

As it is, the term 'projects' was used advisably, bringing the discussion on to the introduction of how conversation analysis is utilised in the analyses. It is exactly a consequence of the interdependent relationship of category work and sequentiality that the moderator is understood as merely *projecting* a certain moral order. Whether or not the moral order is accomplished depends on how the respondents orient to the projection (and, in principle, the moral order is continually open for negotiation). Yet, whereas the respondents can do all sorts of things in response to the moderator, whatever they do will, recalling Sacks, be understood as exactly that: a response to the moderator. The following introduction is aimed specifically at the rather particular sequentiality most evident in the focus group data. Firstly, I provide an introduction to so-called topic-proffering sequences and, secondly, to the peculiarity of yes/no interrogatives.

Topic-proffering sequences are important in this context, as one could say that if a multi second pair parts design constitutes one dimension of accomplishing a focus group context, a topic-proffering design constitutes another.[9]

9 Of course, it is possible to imagine settings that come off as focus group settings, but which do so drawing on other types of sequential resources. Concretely, whereas my

However, such sequences are characterised by contingencies that are different from the ones found in most other types of sequences. In particular, whereas it is ordinarily the case that *dispreferred* seconds are expansion relevant (due to the fact that they are commonly mitigated and accounted for in different ways), in topic-proffering sequences the sequential environment tends to display a systematic reversal of this ordinary differential expansion relevance of preferred and dispreferred second pair parts (Schegloff, 2007a: 169). In other words, in topic-proffering sequences mainly *preferred* seconds are expanded. Accordingly, it is often necessary to adopt a subtler approach to the preference structure in such sequences than the one taken in many other cases.

Basically, a topic-proffering sequence comprises a first pair part, a topic proffer and a second pair part, a response to this proffer. To understand the preference structure of this kind of sequence, these components should be considered one by one. Beginning with the proffer, it is a first pair part in which a speaker (for instance, the moderator of a focus group) proposes a particular topic (for instance, transportation practices), but without elaborating on it further. By proffering, the speaker merely makes the topic available for the recipients (for instance, participants in a focus group setting) to embrace or decline (169–70). Accordingly, the first important feature of topic proffers is that they normally put forward recipient-oriented topics. In particular, recipients are treated as the *proper* speakers on topic, and, by implication, they are likely to carry the burden of talking in topic-proffering sequences. This is due to the fact that the topic most often concerns "something which is specifically, differentially, or even exclusively within the recipient's experience, or on which their view has special weight or authority" (170). Furthermore, a second feature of topic proffers is that they are often, but not always, implemented by so-called yes-no interrogatives.

Reviewing also the second pair parts to such proffers, it becomes evident why topic-proffering sequences tend to have an uncommon preference structure. Since topic proffers commonly project a particular topic, it is readily seen that a *preferred* second pair part would encourage this topic, while a *dispreferred* second would discourage it. However, as Schegloff points out, several features of the response turn are relevant for figuring out its stance towards the proposed topic, that is, for figuring out whether the second pair part is a

data displays an extended use of multi second pair parts, other focus groups have a different and more differentiated dispersion of first and second speakers; that is, in other groups it might be much more common for participants to launch first pair parts and for these to be oriented specifically to other participants, thereby drawing on a format that does not generate multi second pair parts.

preferred acceptance or a dispreferred deselection of the topic. For instance, he mentions that one has to consider the positive or negative response tokens of the response and, relatedly, whether or not it is aligned with the polarity of the question implementing the proffer. Moreover, it is often significant whether or not the response turn is constructed as minimal or not, and, as should be clear by now, in topic-proffering sequences, the minimal response is ordinarily used to display *dis*preference. Moreover, Schegloff observes that the features do not work mechanically (or, I would like to add, independently), but should be seen as a range of ways in which the recipient of a proffer can display his or her stance towards it. In this process, Schegloff suggests, it is of overall importance whether the recipient displays *access* to the topic or denies access to it (171). Accordingly, it should be noted that even though topic-proffering sequences often display an expansion-relevant design that is the opposite of the design of 'normal' sequences, it is not to be understood as completely or necessarily opposite. In particular, and as indicated in the quote, even though preferred second pair parts are expected to be expanded, this does not imply that dispreferred second pair parts are never expanded. Consequently, it should be borne in mind that with regard to topic-proffering sequences, close attention to the subtleties of preference is needed.

Considering yes/no type interrogatives, Raymond (2003) refers to them as one of the most pervasive practices in social interaction (939). As indicated above, this structure is predominantly found in topic proffers, and, accordingly, the practice often features in the topic-oriented focus group data. According to Raymond, this should come as no surprise, since yes/no type interrogatives are more or less ubiquitous in talk-in-interaction (939). Yet, as Raymond observes, sociologists have mostly overlooked the practice (940), and he sets out to balance this disparity, focusing especially on adjacency pair sequences and the varied organisation of preference/dispreference (942).

In this context, it needs to be acknowledged that alignment in yes/no type interrogatives involves (at least) two aspects. In particular, Raymond argues that while alignment is of course a matter of whether a second pair part displays a preferred response concerning the action type generated by the interrogative in question, it should be noted that alignment is also a matter of whether the response is aligned with the polarity of the interrogative (943). Following this observation, Raymond suggests that compared to adjacency pairs generated by other grammatical resources, grammatically formed yes/no type interrogatives mobilise a distinct set of constraints (944). Consequently, responses can conform to these constraints or not, but as Raymond points out, type-conforming responses mainly come off as preferred, while nonconforming responses typically come off as dispreferred and as engendering expansion (947–54). However, it should be observed that this preference is an interactional product, that is, an accomplishment by first as well as second pair part speakers (955).

Accordingly, yes/no type interrogatives are a fine-grained matrix for negotiations of all sorts, displaying highly moral implications. Perhaps it is only necessary to consider a simple, humorous trick question like 'Are you sober yet?' to realise the moral weight of interrogatives. The normative call for a type-conforming answer almost forces the respondent to accept whatever presuppositions the interrogative implies. Yet, as Raymond also points out, first pair part speakers generally design their interrogatives to permit type-conforming alignment, due to the mere possibility that the respondent can, nevertheless, respond in a nonconforming way, and, similarly, that people in general regard questions like 'Are you sober yet?' as humorous. Discussing the extended use of interrogatives in different institutional settings, Raymond exemplifies his understanding of such moral implications: "in questioning a middle-aged woman whose daughter was born around 1960 a doctor asks: 'Are you married?' The doctor formulates this FPP to prefer a 'yes,' thereby revealing his understanding that, for this recipient, being married is preferred over other options" (961). Following Raymond's argument of the moral implications of yes/no type interrogatives, my use of focus groups is further justified as a feasible choice for studying the in situ accomplishment and negotiation of moral order or, I would like to suggest, for studying the in situ accomplishment and negotiation of rationalities of government or the conduct of conduct. I would like to conclude this section by adding two additional perspectives to Raymond's work. The first is related to the particularities of questions in Danish. The second is related to a recent development which clarifies the notion of preference in a way that seems to be well-suited for conceptualising Raymond's observations concerning the different types of more and less preferred responses to yes/no interrogatives.

Firstly, then, I would like to mention two papers by Heinemann (2005 and 2010). In the first paper, "Where grammar and interaction meet: The preference for matched polarity in responsive turns in Danish", Heinemann (2005) suggests that the relationship between grammatical format and interactional preference is if not more prominent in Danish than in English, then definitely not less. In more detail, she points out that there is a strong connection between, on the one hand, grammar and polarity and, on the other hand, interactional preference. This is demonstrated in the overwhelming tendency to mirror the polarity of a prior turn in the response. Moreover, Danish has an 'extra' positive response particle for displaying this preference, namely the Danish 'jo': "this grammatical preference for mirrored polarity in Danish is so strong that a 'special' positive response particle jo 'yes2' rather than ja 'yes1' is deployed in orientation to this pattern when negatively framed utterances, for whatever reason, receive a positive response" (375).

Further, Heinemann (2010) accounts for another type of yes/no question, namely declarative questions. Her observation is important because this type of question is used in my data, and because it emphasises that this form of yes/no questioning invokes the same moral implications as yes/no interrogatives; for instance, declarative questions also pose the restrictions/opportunities related to type and non-type conforming answers (2722–3). Briefly reviewing Heinemann's observations concerning declaratives, she points out that such questions make up the second largest group of yes/no questions in Danish (2705) and that they are "utterances that have no syntactic features that mark them as questions, but are nevertheless treated as such by the participants in an interaction" (2705). Furthermore, she observes that since approximately fifty per cent of such Danish questions lack any prosodic features that mark them as questions (they rarely have an observable rising intonation that can provide for a demarcation from declarative utterances used as different types of actions; for instance, for doing observations), participants most often rely on other features to decide whether a declarative should be treated as a question or not (2706). Heinemann suggests that such features are often of a certain epistemic kind (2706). This is observable in my focus group data where declarative questions are often part and parcel of 'doing being members' of a focus group setting, that is, the participants ordinarily approach the moderator as moderator and, by implication, typically treat her declarative utterances as questions aimed at eliciting more specific information on a certain topic.

Turning to the second perspective, Stivers, Mondada and Steensig (2011) suggest the clarifying concepts 'alignment' and 'affiliation' for the two main forms of cooperation through which relationships, moment by moment, are affected by different social norms. I draw on and consider alignment and affiliation several times in the analysis proper, but for now I would merely like to suggest that it provides for a subtler treatment of, for instance, non-type conforming responses as often dis-affiliated, even though they might be more aligned than dis-aligned concerning, for instance, polarity.

6. The conduct of transportation conduct

The excerpts analysed in Chapter Six and Chapter Seven differ in a rather straightforward way. Whereas the parties to the focus group setting make the municipal strategy observably relevant in the excerpts in Chapter Seven, they do not do so in Chapter Six. Following this difference, the analyses in Chapter Six focus on how one's own transportation conduct is made rationally accountable. On the other hand, the analyses in Chapter Seven focus on how the participants accomplish and negotiate the municipal strategy as a rational attempt to conduct the citizens' transportation conduct. However, a very important point should be made concerning the different foci of the respective chapters. I understand the attempt to make oneself accountable for one's own actions as an inherent part of governing, and, recalling Rose (1999), I consider this account work a part of governing, also when it is undertaken at the most molecular level. Accordingly, I understand the account work in both Chapter Six and Chapter Seven as an inherent part of governing.

Considering Chapter Six in more detail, it demonstrates how the parties to the focus group setting accomplish the rationally accountable moral order of transportation practices. As should be clear by now, I understand the in situ accomplishment of, for instance, an *obligation* to go by car as part and parcel of accomplishing the rational, moral order of transportation practices, that is, of accomplishing the conduct of one's own transportation conduct. Hence, the analyses consist of the very simple job of pointing out how the parties to the focus group setting utilise the machinery of conversation to accomplish their transportation practices as rationally accountable. Yet, as it will be made clear, pointing out how a 'car owner' category is in fact made observably relevant in the process of invoking and utilising its bound obligation to go by car in an account of not using public transportation is no simple undertaking. Accordingly, if the preceding chapters have led to the fairly easy-sounding theoretical argument that the rational accountability of the conduct of automobile conduct is quite simply a matter of *how* people accomplish the rational, moral order of such conduct in talk-in-interaction, the proceeding chapters will show that, in practice, it takes quite a few steps to emphasise such *hows*.

Biking, freedom and fulfilment

In this section I provide an analysis of how biking is approached as something that is inherently good to do but as something that none of the participants have a *primary* obligation to do as a means of transportation. The data stems from a session conducted before the municipal theme month, and it concerned the participants' everyday transportation practices; the moderator asked questions about walking, biking, going by car and using public transportation. The interaction represented in Excerpt One is almost sixteen minutes into the session. At that point, the parties to the setting have stopped talking about walking and have been talking for a while about bicycling.

Excerpt One

Participants: Moderator: mod. Tine: tin. Julie: jul. Berit: ber. Sussi: sus. Also present: Pernille

1	jul	jeg ka' godt li' å cykle i like riding my bike	
2		(.5)	
3		hvis det i:kk' regner s'fø'lig if it doesn't rain of course	
4		[((laughs))	
5	mod	[((shifts gaze from jul to the others, lowered head, lightly	
6		furrowed brow))	
7		[er- er der nogen af jer [do- do any of you	
8		der cykler i regnvejr ride your bicycles when it rains	
9		(.6)	
10	tin	ja yes	
11	(ber)	=ja =yes	
12	jul	ja jeg ka' da osse godt yes well i also	[li' det [like it
13	tin		[jeg ka' osse [i also
14		godt	
15		[li' det hvis det hvis man er klædt på ti' det [like it if it if you are dressed for it	
16	sus	[man- man ka' jo komme ud for det me(h)n [well it can happen to you bu(h)t	

17		[ha ha	
18	jul	[jaer [yes	
19		(.5)	
20	jul	°det da rigtig °that's of course true	[nok° [enough°
21	sus		[=.hh
22		nej jeg tar ikk' decideret ud å cykle no i don't specifically go for a ride	
23		i regnvejr when it rains	
24	mod	nej no	
25	sus	=det gider jeg ikk' =i don't feel inclined to	[vel (men a'så) [you know (but well)
26	jul		[ikk' hvis det [not if it
27		[regner på forhånd nej. så gør jeg det ikk' [already rains no in that case i don't go	
28	mod	[((looks towards sus))	
29		[å de:r ikk' nogen af jer der har f- [and none of you have f-	
30		((shifts gaze direction and looks towards jul))	
31		altså de:r ikk' nogen af jer der har that is none of you have	
32		(.8)	
33		et eller andet som i altid cykler til one or the other that you always ride to	
34		(.3)	
35		å hvor det er lidt lige meget hvad vejret and where it doesn't really matter what the weather	
36		er s-så >a'så de:r ikk' nogen der- is t-then >that is nobody is-	
37		((gesticulates)) (1.9)	
38		f-for eksempel mig som altid cykler til toget. f-for example me who always ride to the train.	
39		(.7)	

40	mod	hvor det [e:r and it [doesn't
41	jul	[uhm [um
42		(.8)
43	mod	lidt lige meget om det regner. >s'fø'lig er really matter if it's raining. >it is of course
44		det ikk' sjovt når det regner not enjoyable when it's raining
45		(.8)
46		men (.) jeg har bar(h)e ikk'(h)ke nogle but (.) i just do(h)n't ha(h)ve any
47		alt(h)ernati(h)ver [.hh al(h)ternati(h)ves [.hh
48	jul	[uhm [um
49	mod	ø:hm å >så'n er der ikk' nogle af jer (.) u:m and this is not something any of you (.)
50	tin	[jeg gjorde det sidste år [i did it last year
51	jul	[( ) hvis de:t [( ) if it's
52		regnvejr jo= raining of course=
53	tin	=jeg var begyndt å =i had started to
54		(1)
55	tin	ø- cykle: 'm det u- ride: 'm my
56		arbejde jeg havde før det var ude på former job it was over at
57		skovcenteret the forest centre
58		(.5)
59		ude [i over [in
60	mod	[ja [yes

61	tin	nakskov de:t- (.4) noget med tretten nakskov i:t's- (.4) about thirteen
62		kilometer kilometres
63	mod	ja yes
64	tin	å der var je:g cyklede jeg var blev and there was i rode i was i
65		jeg rigtig glad for det å investe:rede mig really started to like it and invested me
66		(.) til sidst købt' jeg mig rigtig god cykel (.) eventually i bought myself a really good bike
67		(.3)
68	tin	.hhh
69		(.3)
70	tin	å blev rigtig bidt af det and became really fond of it
71	mod	[ja [yes
72	tin	[a'så] det var noget med å stå [you know] it was about getting
73		tidligt op å komme afsted nogen gange up early and get going sometimes
74		cyklede jeg osse ud å badede tidligt om morgenen i also rode out for a swim early in the morning
75	mod	[ja [yes
76	tin	[å] mødtes med nogen kollegaer
77		[and] met some colleagues
78		(.)
79		jeg syn's det var så, (.) [fedt i thought it was so, (.) [nice
80	jul	[hold da op [oh my
81	mod	ja yes
82	tin	så da jeg fik det nye job her å øh (.) so when i got the new job here and um (.)

83		ka' gå på arbejde i løbet af to am able to walk to work in the time of two	
84		minutter .hh så sagde jeg osse det minutes .hh then i also said that	
85		kunne hun ikk' være bekendt for she didn't have the right to it because	
86		hvad skulle jeg gøre uden min cykeltur ((smiling voice)) what was i supposed to do without my ride ((smiling voice))	
87	All	((laughter))	
88	tin	for de:t ikk' det samme because it's not the same	[å= [to=
89	mod		[ja [yes
90	tin	=cykle en tur det har =just go for a ride i	[jeg= [have=
91	mod		[nej [no
92	tin	=prøvet nogle gange men de:t bare =tried that a couple of times but it's just	
93		ikk' det samme. not the same.	
94	jul	nej no	
95	tin	de:t lækkert å cykle langt på it's great to ride a long way to	
96		arbejde. work.	
97		((gesticulates))	
98		synes jeg å gå i bad å så møde på i think and shower and then clock in at	
99		arbejde work	
100	mod	ja yes	
101		(.)	
102		>fordi de:r et< må:l= because there is a goal=	

103	ber	=et mål ja= =a goal yes=
104	tin	=de:r et mål =there is a goal
105		[>å jeg kan godt li' å cykle stærkt= [and i like to ride fast=
106	jul	[ja det savner jeg osse sommetider= [yes i also miss that sometimes=
107	mod	[((looks towards jul, nods))
108	tin	[jeg blir altid jeg blir (.) [i always get i always get (.)
109	jul	[=(hvis man bare ku' køre) [=(if you could just ride)
110	mod	[((looks towards tin))
111	tin	[svedig jeg skal i bad jeg ka' ikk' [sweaty i have to shower i can't
112		(.4)
113	mod	ja yes
114	tin	jeg ku' ikk'- jeg ka' ikk' cykle i couldn't- i can't ride
115		så'n like this
116		(.9)
117		hvis jeg cykler længe- if i ride long-
118		længere ture så så blir jeg altid longer trips then then i always get
119		svedig for jeg [ka'= sweaty because i [like=
120	mod	[ja [yes
121	tin	=godt li' å (.) få fart på å =i like to (.) speed up and
122	mod	ja yes
123		(.6)
124		hvis man skal cykle tretten if you have to ride thirteen

125		kilometer (.) s- kilometres (.) t-
126	tin	så skal man heller ikk' then you neither
127		[så skal man ikk' (hygge) [then you are not supposed to (relaxe)
128	mod	[det ka' jo næsten ikk' [really it hardly can't
129		lade sig gøre uden å be done without to
130	tin	nej [he-he no [he-he
131	mod	[komme [breaking
132		til å to
133	jul	°nej° no
134	(sus)	°nej° no
135	mod	å svede næsten uanset om a sweat almost no matter if
136	(tin)	nej no
137	mod	om man så forsøgte å cykle langt- if you then attempted to ride slow-
138		det ku' jeg i hvert fald ikk' tror at least i couldn't do that i
139		jeg think
140		(1.8)
141	mod	[øhm [um
142	tin	[da] ku' jeg da godt nyde nogen gange [then] sometimes i could actually enjoy
143		når det regnede å skulle hjem when it rained and was heading home
144		(.4)
145		[.hh

146	mod	[ja [yes
147	tin	efter en lang arbejdsdag å så lige after a long workday and then just
148		det der med å rigtig= this thing about really=
149	jul	=ja =yes
150		(.5)
151	tin	får pulsen op å: blir vå:d å getting the pulse going and: getting we:t and
152		svedig å så hjem å i bad ej så syns sweaty and then home and in the shower oh my then you
153		man lige man er levende igen ikk', just feel you're alive again right
154		(.5)
155	mod	så du tog cyklen til nakskov so you took the bike to nakskov
156		(.4)
157		>osse når det regnede om morgenen also when it was raining in the morning
158	tin	det gjorde jeg i did
159		[(.6)
160		[((mod nods, smiles))
161		ej a'så s'fø'lig no well of course
162		var der nogen dage hvor (.) .hh there were days when (.) .hh
163	mod	ja yes
164	tin	=hvis jeg sku' ti' møde å alle =if i was going to a meeting and every
165		mulige å tænkte ah så tar jeg thing and thought ah then i'll take the
166		[bilen= [the car=

167	mod	[ja [yes	
168	tin	=i dag men jeg har oss cyklet i =today but i have also cycled in	
169		regnvejr rain	
170	mod	ja= yes=	
171	tin	=å nogengange nød jeg det oss =and sometimes i also enjoyed it	
172	mod	ja yes	
173	tin	det kommer (lidt) an på hvad det er it depends a bit on what kind of	
174		for noget slags regn ikk' det der= rain it is right this=	
175	(jul)	=uhm =um	
176		(.)	
177	tin	sjyuw ((downward hand movement))	
178		[((laughs))	
179	mod	[nej som (.) [no like (.)	
180		[som som de seneste dage [like like the last few days	
181	tin	[ne:j (.) de:t ikk'= [no (.) it's not=	
182	jul	[(så: det ikk' så sjovt [(then it's not that funny	
183	tin	=sjovt (.) men så'n =funny (.) but kind of	[stille= [quiet=
184	mod		[nej [no
185	tin	=regn ka' jeg godt li', =rain i do like	
186	mod	s- så m-man=nærmest blevet skyllet væk s- so y-you=almost were washed away	
187	tin	ja yes	

In the interaction represented, one of the participants carefully resists a 'no other alternatives' predicate and, thereby, riding a bicycle is subtly, but explicitly accomplished as a non-obligation. In more detail, the interaction demonstrates how the moderator invokes and draws on a 'no other alternatives' predicate, as she proffers a topic on 'always riding a bicycle, also when it rains'. However, the responding participant, Tine, carefully accomplishes a topic *not* on 'always riding a bicycle, also when it rains', but on the 'enjoyable, fulfilling practice of riding a bicycle, sometimes also when it rains'. Since, as I point out, Tine approaches this accomplishment as doing the job of a completely preferred, topic-encouraging second pair part to the moderator's proffer, and Tine can thereby be seen as resisting the relevance of any 'no other alternatives' predicate. Tine quite simply draws on the affordances of the structural design in order to resist the relevance of a topical context in which such a predicate might be invoked. Hence, the analysis shows how '*always* riding a bicycle, also when it rains' is rendered an irrelevant topic, and how the '*enjoyable* practice of riding a bicycle, *sometimes* also when it rains' is rendered highly relevant, thereby accomplishing riding a bicycle as an occasional enjoyable, fulfilling choice, not as a relevant alternative to automobility.

Considering the most general level of structure of the interaction, the moderator three times attempts to make 'riding a bicycle when it rains' a relevant topic. The first attempt is initiated in line five, the second in line twenty-eight and the third in line 155. As the three attempts are parts of a sequential progression, they of course do different interactional duties The first and second attempts are similar in that they are treated as topic-proffering first pair parts, but they are different in that the first co-constitutes a topic transition and the second co-constitutes a topic development/progression. The third attempt, however, is different in that it does not come off as a proffer, but rather as an ordinary closure-implicative third turn 'formulation'. I have followed the structure provided by these three attempts by dividing the following analysis into three parts, reflecting roughly the interaction generated by the three attempts. Further, the lengthiest part of the interaction is constituted by the second attempt and the response to it. Accordingly, I have again divided this part into two subparts, namely into the proffer and the second pair part, respectively.

Following basic observations of structure of talk in interaction, speakers are specifically accountable for designing their utterances for the recipient,[1] and, as Raymond (2003) observes, this of course includes yes/no interrogatives. In

1 Cf. Garfinkel, 1967; Goodwin, 1980 and 1981; Sacks and Schegloff, 1979; Sacks et al., 1974; Schegloff, 1972.

particular, it should be recalled that he observes that when speakers draw on this form of interrogative, they are accountable for designing their utterances in a way that appropriately reflects what the parties to the interaction all know and what can be presumed concerning the matters enquired into, for instance, as a consequence of what has or has not been established by prior talk. As discussed, this is a consequence of the specifically morally implicative features of such formats. Similarly, it should be recalled that whereas Raymond is concerned in particular with grammatically formed yes/no interrogatives, Heinemann (2010) shows that Raymond's observations are applicable to grammatical interrogatives as well as to declarative questions in Danish (2722–3).

Considering that it is part and parcel of 'doing being parties to a focus group setting' that the moderator asks questions, it is not too surprising that the moderator uses different forms of yes/no questions in each of her three attempts to make 'riding a bicycle when it rains' a relevant topic. Similarly, as Schegloff observes (2007a: 170), topic proffers are most often implemented by such designed utterances. Accordingly, it can be seen as 'doing being parties to a topic talk-oriented focus group setting' when the moderator draws on yes/no questions and the participants fulfil these proffers in extended second pair parts. Hence, the interaction is co-constituted by the affordances and limitations of, on the one hand, the 'doing being parties to a focus group setting' and, on the other hand, the moral implications of the yes/no questions. In the present excerpt this co-constitution provides for particularly subtle interactional negotiations. Accordingly, the excerpt can be seen as a refined demonstration that sequence and category are two sides of the same coin, in that the negotiating category work is notably hinged on the structural affordances.

As mentioned, I am particularly interested in how Tine resists the 'no other alternatives' predicate invoked by the moderator. Since the moderator invokes the predicate in her second proffer, and Tine negotiates it in the following, I am particularly interested in the interaction initiated in line twenty-eight. However, since the second proffer is closely intertwined with the preceding interaction, it is necessary to refer back to these lines to describe what is going on from line twenty-eight and onwards, and it is, accordingly, useful to begin with a more thorough overview of the first lines and the first proffer displayed in the excerpt.

Before the excerpt, the interaction unfolded in a fairly long topic-proffering sequence with multiple second pair parts to the moderator's proffer concerning the participants' biking practices. In line one of the excerpt, Julie concludes her second pair part by the summarising, "I like riding my bike". Then follows a short pause, and in line three Julie adds that it is of course only the case when it does not rain. These lines are closure implicative in that they project what Svennevig (2000) refers to as a 'topic transition relevant place', namely a "recognizable [point] in a conversation where the interactants display the potential for closing a topic and introducing a new one" (188). Furthermore, Svennevig notes that topic closure is projected through a certain lack of progressivity, which takes the form not only of silence, but also of specific types of turns: "Types of turns that do not contribute to developing the topic are those that are predominantly responsive, such as minimal responses, repetitions, reformulations, summaries, generalizations, conclusions and the like" (188). Hence, topic closure is projected by reactive utterances that serve to put into perspective previously presented information rather than to add new information. Accordingly, since Julie's summarising assessment in line one comes off as characterising the biking she has been reporting on, it comes off as projecting a topic transition relevant place.

Notably, Julie's closure projection is not followed up. Nobody, neither the moderator nor the participants, self-selects and delivers an aligning closure implicative acknowledgement of Julie's assessment in the pause in line two. Then, in line three, Julie continues, weakening her assessment and displaying a reach for alignment. Similarly, Julie's laughter in line four is an example of her reach for alignment, as it underlines the weakening of her assessment displayed via the 'selvfølgelig'/'of course' in line three. Yet, when the moderator responds, producing her turn in lines seven and eight, she does not collaborate on the topic of closing. Instead the moderator turns to the other participants and asks whether any of them ride their bicycles in the rain, thereby displaying a new focus within the same topic. Consequently, the moderator seems to *restore* the progressivity of the topical talk, but it should be noted, however, that this is carefully done, acutely orienting to the preceding interaction. This should be considered in a number of steps.

In lines five to eight the moderator shifts her gaze, looking away from Julie and instead focuses her eyes on the other participants. Furthermore, she lowers her head and furrows her brow. She displays a 'doubting face' when she delivers her proffer, implemented by the yes/no interrogative 'er der nogen af jer der cykler i regnvejr'/'do any of you ride your bicycles when it rains'. Also, she does not emphasise the 'doing questioning' with an upward intonation; on the contrary, she stresses 'nogen'/'any', thereby stressing an 'at least one', rather than a 'perhaps all' hearing. Lastly, it should be noted that the 'doubt' is also reflected in the repair of 'er'/'do'.

The moderator's response does a number of things in the interaction concerned. As stated, the response does not straightforwardly relate to Julie's preceding assessment, since it does not directly align with it (for instance, through a typical acknowledgement token like 'of course not' or 'no, me neither'). However, since the response is non-verbally oriented to everyone but Julie, it displays a context in which Julie has already answered the question and, thus, an understanding of Julie's assessment as including such an answer. Specifically, the moderator's response treats Julie's negative evaluation of biking when it rains as doing the job of accounting for the fact that she does not do so, which, it should be noted, Julie has not said. Yet, the audible stress on 'nogen'/'any' further emphasises that the moderator displays such an understanding of Julie's work, in that it, as indicated, appears to take 'nogen' to literally mean *everyone*. In this way, the moderator's response negotiates what might have been a concluding evaluation, treating it as projecting a possible extension area of the biking topic. Simultaneously, by displaying the topic progression as incorporated in Julie's work, the moderator's 'lacking' topic closing collaboration comes off as less lacking.

Following this reading, I will suggest that the moderator takes the liberty to negotiate a potential topic closure into what I, again with Svennevig (2000), will call a topic transition. Svennevig describes such transitions in the following way: "Gradual topic transitions therefore do not include topic introduction but merely adjustments of already established topics. A joint project may be adjusted under the course of its completion, but the adjustment is not itself a new project" (210). As the moderator treats Julie's preceding work as primarily adjusting the topic so that it can also include situations in which it rains, she treats it as a gradual topic transition, not as an instance of topic closure.

Hence, it can be noted that the moderator is 'doing being moderator', drawing on her rights to ask questions. Furthermore, it can be noted that she does so in the morally implying form of a yes/no interrogative. However, as already touched upon, this would be a decontextualised, insensitive hearing if the moderator was heard to ask a simple question, unambiguously implying that the participants *should* be riding their bicycles when it rains. Rather, it should be noted, the moderator does not in any way disregard or neglect Julie's preceding action, but, in contrast, draws on the opportunity it provides for carrying on the topic. This opportunity, however, hinges on a specific understanding of Julie's utterance that she does not like to ride her bicycle when it rains: an understanding in which 'I do not like to ride my bicycle in the rain' is synonymous with 'I do not ride my bicycle in the rain'. Not only does the moderator suggest such an understanding, she also subscribes to it, both through her emphasis on 'nogen' and through her bodily display of 'doubt'. Hence, even though the grammatical structure of the moderator's request makes 'yes' the type-conforming, preferred response, it can be said to anticipate a negative answer in the way it relates to the preceding interaction. In other words, the moderator seems to walk the thin line between, on the one hand, 'doing being a *right*-full and *obligated* moderator', promoting the topic talk and, on the other hand, aligning her moral implications with what she displays as *her* understanding of an already projected moral order.

Continuing with the participants' responses, it should be noted how carefully they maintain the subtle balance. In more detail, whereas the responses conform to the structural call for a type conforming 'yes', the participants simultaneously further the accomplishment of riding a bicycle as an activity predominantly done if one enjoys it.

Through the 'yeses' in lines ten to twelve the participants fulfil the moderator's request, a topic transition initiator, and in the following lines they embrace the changed topic by claiming to have access to resources that can sustain the talk, not merely on riding a bicycle, but also now on riding a bicycle when it rains. The participants do so as they in lines twelve to twenty-seven negotiate the more specific meaning of the topic transition 'riding a bicycle when it rains', focusing on a matter of 'likeability' (Julie and Tine mention that they 'enjoy' riding their bicycles when it rains in lines twelve and fifteen) and matters of 'happenstance' versus 'planned' biking. The latter is evident when Sussi in line sixteen reveals the understanding that when riding a bicycle, one can of course be surprised by a shower, and when she, in lines twenty-two and twenty-three, and Julie, in lines twenty-six and twenty-seven, say that they do not go for a ride if it is already raining. Whereas Sussi's response does not readily come off as oriented to Julie's utterance in line twelve (and to Tine's agreement with Julie which Sussi is overlapping), it comes off as displaying much the same understanding of the moderator's question as the one displayed through Julie's (and Tine's) response. That is, Sussi's response is also oriented to the moderator's question about whether any of the participants ride their bicycles when it rains as a matter of whether any of them enjoy riding a bicycle when it rains, or, in Sussi's case, *intend* or *plan* to go for a bike ride in the rain. When Sussi says that she 'ikke decideret'/'don't specifically' goes for a ride in the rain, she treats the moderator's question as, for all practical purposes, oriented to riding a bicycle when it rains as a matter of individual disposition. Not, for instance, as a matter of everyday necessity. Hence, in these first few lines on riding a bicycle when it rains, 'enjoy riding a bicycle when it rains' and 'plan not to ride a bicycle when it rains' come off as observably relevant answers to the moderator's question, 'er der nogen af jer der cykler i regnvejr'/'do any of you ride your bicycles when it rains'.

Against this understanding of the first proffer, it is time to consider the ongoing interaction, namely the moderator's second proffer in lines twenty-eight to forty-nine. As mentioned, the second proffer displays an expected attempt to carry on the topic. In more detail, Svennevig (2000: 172) points out that topics are organised around the principles of reportability, projectability, connectedness and progressivity, arguing that it is possible to recognise talk on topic and the organisational status of one or another contribution. The principle of progressivity refers to the fact that topics can only be carried on for as long as they progress, and Svennevig observes that the requirement is applicable both to aspects of form and content; that is, progressivity is a matter both of continuous talk without gaps and of adding new information (184).

Since the moderator's second proffer appears to reach for new information on the topic of riding a bicycle when it rains, it can be seen as a development of the same topic. Hence, again the moderator displays 'doing being a topic *generator*' rather than 'doing being part of the collectivity of participants' who are properly entitled to deliver the information that can carry on the topic. In other words, through the proffering, the moderator not only invokes her own membership of 'moderator to a focus group setting', she also projects the participants' memberships of the collectivity 'participants to a focus group setting'. However, the topic progression is not fulfilled until lines 142 to 143. It is not until these lines that the responding participant, Tine, finally focuses on matters of riding a bicycle when it *rains* and, as it is, that it might look as though the moderator's work is instead accepted as an instance of topic *transition*. Hence, two points can be noted: firstly, the moderator uses quite a few lines to deliver this second attempt to carry on the 'riding a bicycle when it rains' project and, secondly, it takes a remarkably long time before the moderator's attempt is accomplished (and when it is, it is, as demonstrated below, accomplished in a negotiated version).

Considering the moderator's lengthy request in lines twenty-eight to forty-nine, the explicitly proffering lines are (accomplished as)[2] negatively formatted yes/no declarative questions. In lines twenty-nine to thirty-three the moderator roughly asks, 'and none of you have something that you always ride your bicycles to?', and in line forty-nine she says, 'and this is not something any of you do?'. As Heinemann (2005) has shown very clearly, a confirmation of a negative statement takes a negative form (versions of 'nej'/'no'), but as Heinemann also shows, negatively formatted utterances can be designed to go against this general preference, rather than to disconfirm the literal assertion. Heinemann

---

2 This explication is supposed to invoke Heinemann's (2010) observations concerning declarative questions in Danish. In declarative questions in Danish, there are no syntactical and (most often) no prosodic features that display that the declarative is used to do 'questioning', and, by implication, participants have to base their understandings on other features. Hence, in a way, such use of declarative questions forces the analyst to do what should always be done, namely to orient to the elements of observable relevancy in the data.

demonstrates that this calls for a specific version of 'ja' (yes1), namely 'jo' (yes2) (393–4). Hence, recalling Raymond's observations, it can be noted that a 'nej'/'no' would, structurally speaking, be the preferred, type-conforming answer, whereas 'jo'/'yes' would be the dispreferred, type-conforming answer. However, this is only insofar as it concerns only the most superficial structural design. As it is, deciding whether the more aligned response to a particular negatively formatted request is 'yes' or 'no' always requires close analytic attention.

Accordingly, it is necessary to consider how the moderator's second proffer is connected to the preceding interaction to better decide how possible preferences are evoked in the question that unfolds. The moderator starts to speak in line twenty-nine, and at this point both her body and gaze are oriented to Sussi. Because of the physical orientation towards Sussi, this first line (twenty-nine) does not appear to recognise that it overlaps with Julie's lines twenty-six and twenty-seven. However, as the moderator breaks off at the end of line twenty-nine and in line thirty shifts the direction of her gaze and looks to Julie, it appears to acknowledge the overlap. Yet, as the moderator in line thirty-one self-repairs, she more or less repeats the line that overlaps, only reframing her initial 'og'/'and', replacing it with the more explaining 'altså'/'that is'. This repetition shows that the moderator understands her own project as, for all practical purposes, locally appropriate. Considering the moderator's above-mentioned display of 'doing being moderator' in this work, this orientation indicates that she is willing to use the 'power' of that membership to steer the proceeding topic development in a specific direction.

Simultaneously, however, the moderator is subtly oriented to the preceding interaction in these initiating repair lines; that is, from line twenty-eight to the beginning of line thirty-six, where she completes the first wording of her project (the ending is marked by her beginning repetition, 'altså der er ikke nogen'/'that is nobody is' in line thirty-six). Despite of the overlap, the moderator audibly initiates her turn with an 'og'/'and' in line twenty-nine. The conjunctive connection importantly ties her work to the previous work, projecting that the moderator's action is oriented to *adding* information to the topic at hand (this is supported by the moderator's overlap and repairing repetition). Hence, the moderator's first lines display 'a next on the agenda question', seemingly trying to take the views aired by the participants her point of departure for moving the discussion forward. Against this, it should be recalled that nobody has indicated that they are frequently forced to ride their bicycle in the rain. So, based on the knowledge of the participants, for instance, at the first possible completion points in lines thirty-three and thirty-six, the *first* version of the moderator's question is not specifically directed towards disconfirmation (which could be something like, 'Yes, I have to ride my bicycle when I go to dance classes'). This, of course, indicates that the moderator may have understood the participants' responses, namely Julie's and Tine's work on 'likeability' and Sussi's work on 'planning' (and Julie's, but the moderator is not oriented towards it, as they overlap) to imply that they are never in a situation where they do not consider whether it is raining or not. It can thus be noted that the moderator already initially projects talk on practices of riding a bicycle no matter what as a somewhat unlikely topic or, in other words, that the moderator at one and the same time displays orientation towards topic progression and towards the fact that the participants might not have the resources to sustain talk on this progression.

Significantly, the moderator gets no support (specifically at the possible completion points, lines thirty-three to thirty-four and thirty-six to thirty-seven), and now she starts to use herself as an example. The moderator exemplifies the type of practice the proffer is oriented to. In lines thirty-eight to forty-three, she mentions that she has to ride her bicycle to the train station whether it rains or not. Remarkably, in lines forty-three to forty-four, she adds that riding a bicycle when it rains is 'selvfølgelig ikke sjovt'/'of course not enjoyable', and then, in lines forty-six to forty-seven, she says that she has no other alternatives, thereby seemingly delivering an account of doing it all the same. Hence, before proceeding it should be observed that the category work appears to reinvoke the already suggested fixed relation between 'likeability' and 'riding a bicycle', in that it displays the moderator as specifically accountable for doing something which is not enjoyable.

Considering the category work in more detail, the moderator initiates these lines with a 'men'/'but' that exhibits a disjunctive, here accounting, orientation to the preceding.[3] Through this, the moderator invokes and subscribes to a category-bound 'no other alternatives' predicate as an excuse or justification resource. To appreciate this suggestion, it should be reconsidered what the data reveals. Hence, it has been made observably relevant that the moderator always rides her bicycle to the train station and that she has no other alternatives. This probably invokes the hearing that the moderator is a 'train commuter' and, perhaps, by extension, that she is 'an employee', 'a student', 'a client' or belongs to another category whose members are obligated to go somewhere. Furthermore, it might provide for the hearing that train commuters have to get to the train station because they have to take the train, and that this particular commuter, the moderator, lives at a distance from the train station which makes riding a bicycle not only an option, but the

3 This piece of category work is the one I drew on when exemplifying the subtleties of membership categorisation analysis in the methodology chapter.

only option. That is, the data provides for the hearing that the moderator thinks that she does not have access to, for instance, a car or public transportation, or that she does not have the skills required for either of them. In short, the moderator is likely to be heard as doing a self-identifying appropriation of certain access and/or skills.

However, even as this comes off as a plausible hearing, it is still necessary to show that it is actually the hearing made observably relevant *in the data*. Here, it is helpful to recall Sacks's (1992a) observation that a common part of our everyday orientation and solution to problematic utterances is to assign a category to the person who performs the utterance; for example, the category of 'people who are entitled to say such a thing' (180). Against this, it can be noted that the 'no alternatives' predicate is doing 'justification', since it is invoked to explain what the moderator *collaboratively* treats as a problematic activity, namely her production of the utterances, 'I ride a bicycle when it rains' and 'I do not enjoy riding a bicycle when it rains'. In particular, it can be seen that the purpose of the 'no alternatives' predicate is to establish that the moderator belongs to a category of persons who are entitled to say such a thing. It can also be seen that the specific character of the category to which the moderator 'belongs', for example 'employee' or 'student', is not relevant for the interactional task at hand. Hence, it can be specified that the moderator subscribes to a self-identification regarding certain access/skills and that she thereby invokes a binding category incumbency, but it *cannot* and need not be specified which incumbency we are talking about. I would like to stress the job the predicate is doing. Specifically, by treating 'I ride my bicycle when it rains, even though I do not enjoy it' as contradictory, the moderator's self-identification work comes off as projecting the accomplishment of a moral order in which it is understood as observably accountable to ride a bicycle when it rains, even if it is not considered enjoyable.[4]

Against this category work, the moderator ends her proffer, repeating the (accomplished as) negatively formatted yes/no declarative question. That is, in line forty-nine she says, 'og sådan er der ikke nogle af jer'/'and this is not something any of you'. As argued, the *first* version of this negatively formatted question did not come off as specifically directed towards disconfirmation. Since then, the moderator has projected a moral order in which it is accountable to perform the activity she asks about. Accordingly, this second version of

4 Furthermore, there are audible laughter particles in the moderator's voice as she delivers the account. The laughter, of course, supports the understanding that the moderator treats her activity as contradictory and her 'no alternatives' predicate as a mitigating account.

the question might be even less oriented to disconfirmation. In more detail, in choosing a type-conforming disconfirmation of the inbuilt assumption (by saying 'jo' [yes2], 'you are wrong to assume that none of us have something we need to ride our bicycles to') the respondents would align with the moral description and depict themselves as members of the same category as the moderator; that is, a category which has been described as morally accountable. Contrastingly, in choosing a type-conforming confirmation (by saying 'nej' [no] 'you are right to assume that none of us have something we need to ride our bicycles to'), the respondents would resist being ascribed to morally accountable membership. However, a type-conforming confirmation might jeopardise the respondents' membership of 'doing being parties to a focus group setting'. That is, disregarding the negatively formatted and morally implying request, the moderator's turn can still be heard as doing proffering and as such as projecting some form of topic encouragement as a preferred response. Hence, the moderator's second proffer seems to come off at the very thin line between, on the one hand, the obligations and rights of the 'doing being parties to a focus group setting' and, on the other, the moral implications of her yes/no interrogatives.

Remarkably, the responding participant accomplishes this fine balance by providing a non-conforming answer; that is, in line fifty Tine says, 'jeg gjorde det sidste år'/'i did it last year'.[5] Again revisiting Raymond (2003), giving a type non-conforming answer seems to be a way of going against some inbuilt presupposition in the question, formulating an answer that is more independent of the way the question was put. In more detail, by opening a story with a past biking experience Tine seemingly resists the morally implicative membership of 'persons who ride a bicycle even when it is not enjoyable', but ascribes to a membership of parties to a focus group setting. Before going into more detail with Tine's response, it might be useful to consider the distinction between alignment and affiliation, suggested by Stivers, Mondada and Steensig (2011). As touched upon, they suggest a distinction between *alignment* at the structural level of cooperation and *affiliation* at the affective or action-oriented level. Considering their overview of the features of alignment and affiliation it might be possible to say that Tine's response is disaligning, but affiliating. Since Tine does not give the type-conforming answer 'no', the response is readily seen as disaligned with the formal design preference of the moderator's turn, and, as Raymond observes,

5 It should be noted that Julie also initiates an answer in line fifty-one or fifty-two which is, as it is, partially inaudible. Accordingly, I could not say for sure that Julie does not initiate her answer with a 'jo' (yes2), but I can say for sure that it is Tine's answer that is carried on and, thus, her response is the one who is made observably relevant.

this might be part of going against some inbuilt presupposition in the first pair part. On the other hand, as Tine's response seems to be cooperative concerning the topic talk activity, it seems to be affiliated at the affective level of cooperation. However, it is much more difficult to determine how the response fits with the remaining features. In fact, it might even be that it aligns with the two first alignment features and disaffiliates with the two first affiliation features. Considering, firstly, the two alignment features, Schegloff's observations on the peculiarities of topic-proffering sequences should be recalled. Schegloff observes that it is often a more subtle matter to establish whether a second pair part is aligned or not with a topic-proffering first. Hence, it might very well be that Tine encourages the topic-proffering sequence, taking on her interactional role as one of the proper speakers on the topic, but, at the same time, that she resists some of the presuppositions and terms, going against the formal design preference. In other words, it might simply be that topic-proffering sequences represent a form of activity that does not fit easily with the proposed distinction. Considering, secondly, the first two affiliation features, the issue becomes more complicated, but it might, however, still be related to the peculiarities of topic-proffering sequences and to the certain restrictions and opportunities afforded by yes/no interrogatives. In more detail, I would want to suggest that what happens is that Tine utilises the interactional power of being a second pair part speaker in a topic-proffering sequence to negotiate the topic at hand; thereby, she avoids considering displaying empathy or endorsing the stance displayed in the moderator's first pair part. Seeing as I have not discussed Tine's lengthy response yet, it is of course not possible at this point to agree with my suggested understanding of how it fits into and challenges Stivers, Mondada and Steensig's distinction. However, I hope that my suggestion will provide a reference point that makes it easier to follow the discussion of Tine's response.

Tine's response comes off as an extended narrative of a personal experience of a previous 'riding her bicycle to work' practice. Hence, before getting into the finer details of the response it might be a good idea to recall the most overall points, namely that Tine carefully accomplishes a topic *not* on 'always riding a bicycle, also when it rains', but on the 'enjoyable practice of riding a bicycle, sometimes also when it rains', thereby resisting the relevance of any 'no other alternatives' predicate. It is part and parcel of this work that Tine meticulously draws on the affordances of the structural design to accomplish this work; that is to say, by displaying her own response as a perfectly preferred second pair part, she simply resists the relevance of a topical context in which such a predicate might be invoked. In line with this, the fact that Tine does not mention rain until lines 142 and 143 is quite striking, even on the more superficial level.

However, taking one step at a time, the first thing to note is that Tine and Julie present their responses as perfectly preferred responses. Specifically, it can be noted that even though the moderator does not select a next speaker (but rather approaches the participants as a collective), there is no delay and no hedging markers before the response. Oppositely, Tine and Julie both self-select and co-constitute an initially overlapping response which is partly inaudible for Julie's part.[6] Hence, this unambiguous display of preference supports the understanding that Tine's response should be seen as affiliated with the moderator's project, regardless of the fact that it does not match the formal design preference.

Furthermore, the affiliation is evident from Tine's choice of response, namely in the fact that she delivers an extended narrative on a personal experience of riding a bicycle. However, from the very beginning, Tine utilises the power provided by responding to a topic proffer to negotiate the projection of that very proffer. The way she presents her narrative displays that she understands the moderator's proffer as first and foremost aimed at eliciting accounts about biking *experiences*, not necessarily riding a bicycle in the rain. To appreciate the fine details of this interactional move it is useful to consider Heritage's (2011) work on the differences between knowledge and experience. Heritage's point is that experience is treated differently than knowledge in interaction in that experience is treated as more private and inaccessible than knowledge. Accordingly, as Heritage points out, affiliating with another person's experience constitutes a different kind of problem than affiliating with more readily shareable knowledge. He argues that 'parallel assessments' represent one way of getting around such interactional difficulties of sharing individual experiences. In more detail, he suggests that one form of affiliation is via responses which report similar, though not identical, experiences (168). Following Heritage, it is possible to understand Tine's response to the moderator's work as an instance of 'parallel assessment'. When Tine says, 'jeg gjorde det sidste år'/'i did it last year', it comes off as affiliating with the moderator's proffer, thereby treating it as first and foremost a proffer for experiences of biking. Tine's shift from present to past tense is the first display of her orientation to doing a 'parallel assessment' of an experience. That is, considering that the moderator, on a strictly structural level, could be heard as asking about the participants' present practices, the past tense comes off as a prominent feature that underlines a negotiated orientation, not to (im)possible present practices, but to reports of (necessarily past) experiences. Additionally, in the proceeding interaction, Tine provides an extended narrative of her experiences of riding her

6 Even though it is inaudible, it is definitely *not* delayed.

bicycle to work in her former job, thereby supporting the understanding that she is predominantly oriented towards experiences of riding a bicycle on an everyday basis. Initially, two important and interrelated points concerning this stretch of interaction, this parallel assessment, should be noted.

Firstly, Heritage suggests that members who deliver parallel assessments ordinarily find themselves in emergent 'dilemmas of affiliation', where they not only have to consider the risk of being heard as flat, but also the risk of being heard as "competitive with the very report that it is designed to affiliate with" (169). However, I would like to suggest that this is probably not a main dilemma in topic-proffering sequences; that is, in environments where the recipient is treated as the properly knowledgeable or authoritative speaker. In more detail, I would like to suggest that second pair part providers in topic-proffering sequences have the right and obligation to 'overbid' the report they affiliate with. Accordingly, I would like to suggest that Tine's extended response displays an expected orientation to the sequence as a topic-proffering sequence in which Tine is in the business of delivering a perfectly preferred topic-encouraging second pair part. Hence, in providing her lengthy narrative, Tine ascribes to a membership of 'properly knowledgeable on the topic', thereby invoking her membership of 'party to a focus group setting'.

Secondly, this focus group 'membershipping' is exactly what provides Tine with the opportunity to resist the 'membershipping' invoked in the moderator's first pair part. It is by drawing on her membership of 'properly knowledgeable on the topic' that Tine can provide an account of the past practice of riding a bicycle to work every day as specifically *enjoyable*. In this way, Tine can be understood as drawing on her cooperative affiliation in order to negotiate the projected category-bound predicate; that is, she can be understood to draw on the affordances provided by her affiliation to resist one of the aspects of alignment, namely categorial presuppositions. Bearing this in mind, it is time to go into some of the features of the first part of Tine's extended second pair part.

Tine begins by delivering a typical preface to her biking narrative, setting the scene in lines fifty-five to seventy. Firstly, she points out what time she is talking about, namely the time of her former job (line fifty-six). Secondly, she provides some interesting category work as she points out the place. Here it is useful to recall one of the first papers that build on Sacks's work on categorisation practices, namely Schegloff's 1972 paper in which he draws on analysis of sequences in which members use place to do interactional work. The work leads Schegloff (1972) to the following observation on 'location categories': "For any location to which reference is made, there is a set of terms each of which, by a correspondence test, is a correct way to refer to it. On any actual occasion of use, however, not any member of the set is 'right'" (81). Considering this observation of the interactional use of place, it should be observed how Tine refers to her former workplace. In line fifty-seven, Tine simply names the place 'skovcenteret' (roughly, in English, this would be 'the clinic in the woods'). Then, after a short pause, Tine specifies the location, using the location-referring preposition 'ude i'/'over in'. Notably, the moderator already displays her understanding of this reference when she overlaps with Tine's 'i', certainly *before* Tine pronounces the specific town name 'Nakskov' (line sixty-one). However, notwithstanding the moderator's display of understanding and the fact that everybody present (and everybody on the island of Lolland, I am sure) knows the distance between Nakskov and Horslunde, Tine specifies her location reference further, as she in lines sixty-one to sixty-two mentions that it is 'about thirteen kilometres'. For all practical purposes, Tine thereby displays *distance* as of specific relevance.

Lastly, the preface also includes a personal evaluation of the experience of this former practice of riding her bicycle to work. In lines sixty-four and sixty-five Tine points out that she started to enjoy it, and in the last line of the preface (line seventy) she repeats the evaluation with a slight reformulation, saying that she became really fond of it. Hence, already in the preface, Tine suggests that riding a bicycle to work is really about enjoying the experience of doing so. This suggestion is further developed in her following work.

Hence, in line seventy-two Tine begins her characterisation of the experience itself, marked by the explaining initiator 'altså'/'you know' and by the increased amount of detail in the report. For instance, in lines seventy-two to seventy-nine, Tine provides an account of her biking practice in general, followed up by a more specific example of bathing at the beach in connection with this practice. It should be noted that Tine draws on this concrete example to provide yet another positive assessment, namely the evaluation 'det var så fedt'/'it was so nice' (line seventy-nine). The anaphoric pro-term 'det'/'it' in line seventy-nine comes off as referring to the just described bathing experience. In other words, Tine draws on her detailed 'detour', the more particular bathing experience, to provide yet another *positive* parallel assessment of biking practices. Not only does the assessment come off as pointedly upgraded through the stressed adverb 'så'/'so', it can also be seen as reached for specifically through the bathing story.

Furthermore, it is again useful to consider Heritage's (2011) work on the subtle and not easily fulfilled moral obligations in relation to reports on experiences. Concluding on observations from data, Heritage suggests that there seems to be a certain paradox concerning receiver obligations. Firstly, he suggests that while it seems to be relatively easy to respond properly to reports on more general experiences, it does not seem to be that important to do so, and, secondly, that it seems to be fairly difficult to display alignment with highly granulated accounts of more specific or specialised experiences, although receivers seem to be highly obligated to do so (181). Thus, according to Heritage, it is not that surprising that narratives and reported speech often play a major role in reports on experiences, since these types of accounts "take recipients 'into the moment' and supply the resources for empathic response" (177).

Accordingly, Tine's use of a narrative design can be seen as reaching towards affiliation. It can be noted that Tine succeeds in reaching an empathic moment when Julie provides an into the moment response with her 'hold da op'/'oh my' in line eighty. Heritage remarks that such non-propositional responses represent the only way that recipients can display full commitment to a reported experience. Specifically, he observes that such 'response cries' have a great interactional advantage in that they do not need to discriminate between teller and recipient feelings: "By not discriminating between feelings that the teller associated with the event, and the sentiments the telling is arousing in the respondent, response cries evoke and claim a degree of empathic union and affiliation between teller and recipient" (176). Following this, I would like to suggest that by fulfilling Tine's projection of the empathic moment, Julie could be heard to support or take part in the resistance of the 'no alternatives' predicate projected by the moderator.

Lastly, Tine draws on her assessment and the empathetic moment as a legitimation resource. In more detail, in lines eighty-two to eighty-six, Tine draws on the cooperation as an explanation of how she complained to her employer when she was transferred to her new job only two minutes from her home and thus did not have to ride her bicycle to work anymore. Furthermore, in line eighty-eight, this is supported with an accounting assessment. In particular, it should be noted that in the following lines Tine delivers the assessment that it is not 'the same', just going for a ride in her free time, thereby orienting to her new job as presenting some sort of *obstacle* to her practice of riding a bicycle. In lines ninety and ninety-two, she supports the account by mentioning that she herself has *experienced* its validity. Hence, in the account, it is as if Tine finds that her former job afforded a specific *opportunity* bound by the locational category of the workplace. Accordingly, Tine further negotiates the 'no alternatives' predicate; riding her bicycle to work is not a matter of having no alternatives, but a matter of having the *opportunity* to do so. In this way, biking to work is made relevant as an attractive opportunity, *not* as a 'not enjoyable' necessity.

By providing an overall assessment of the biking to work narrative, Tine sums up the narrative of this past experience in lines ninety-five to ninety-nine. In line ninety-five she says that she finds it 'lækkert'/'great' to ride a bike to work, shower and then 'clock in'. In this way, the assessment is not oriented to any form of biking, but to the specific biking practice that Tine has just accounted for. Hence, since Tine's assessment comes off as grounded in the fine-grained interaction that precedes it, it comes off as if Tine has a certain right to provide the assessment. In more detail, it is worth noting Stivers, Mondada and Steensig's (2011) observation concerning the proper grounding of assessments. They point out that there "appears to be a norm that speakers should make assertions only when they have sufficient knowledge and rights to do so and that speakers with more detailed and in-depth knowledge have primary rights to make assertions and assessments regarding this domain" (14). Following this, Tine's assessment arguably comes off as delivered by an observably and specifically knowledgeable member, and, by implication, it strengthens Tine's resistance to the 'not enjoyable, but no alternatives' predicate.

Before proceeding to Tine's final mention of rain, it should be noted that Tine increasingly orients towards the joy of riding a bicycle as a matter of the *embodiment* of the practice. In line 105 she elaborates on her positive orientation to riding a bicycle by underlining its embodied character. She stresses that she likes riding a bicycle, because she enjoys biking fast, getting sweaty and showering before starting work. Also, just before and as part of this embodied account, Tine mentioned that she likes to bike *long* distances (line ninety-five). In this way, riding is made relevant not only as an attractive opportunity, but, more specifically, as an embodied practice through which Tine fulfils herself as a strong, healthy and enduring individual.

After the delicate accomplishment of biking as an enjoyable and embodied practice, Tine finally addresses rain in line 142, and, notably, she specifically looks for an opportunity to do so. Tine's turn in 142 is initiated with an audibly stressed 'da'/'then' which comes off as a reference to the past she has already thoroughly described. By projecting that the coming turn is placed within 'narrated time', the use of 'da'/'then' accomplishes the moderator's immediately preceding work as being merely a side sequence.[7] While Tine could have addressed the moderator's work as, for instance, some sort of topic leap (Svennevig 2000: 201), she instead draws on her self-provided next speaker affordances in order to render the moderator's work a mere parenthesis. The dismissal is done on the backdrop of a setting in which the dismissed part is otherwise oriented to as the moderator of a focus group setting, thereby emphasising the dismissal. Hence, Tine's 'da'/'then' importantly projects the specific *relevancy* of the proceeding

7 The term side sequence was proposed by Jefferson (1972) to cover a general category of insertions into an established activity (294).

interaction. This is underlined by the fact that Tine's 'da'/'then' almost overlaps with the moderator's (next speaker indicating) 'øhm'/'um' in line 141: Tine can be heard as specifically reaching for the next speaker privilege, which means, importantly, that she can be heard as specifically reaching for her final orientation towards riding a bicycle when it rains.

The orientation to rain comes off through yet another account of an enjoyable embodied experience. In lines 142–3 and 147, Tine says that she 'could sometimes actually enjoy when it rained when she was heading home'. In the local context, the modal adverb 'da'/'actually' invokes a certain 'opposite-ness' or 'contrary to what some might have thought'-ness and, as such, it displays a particular orientation to the moderator's negative evaluation of biking in the rain. Furthermore, the modal verb 'kunne'/'could' invokes an aspect of *ability*; that is, it invokes a form of self-identification as someone who is '*able* to enjoy'. Thus, Tine draws on her particular cognitive dispositions as a resource for resisting the moderator's negative evaluation through the production of a positive evaluation of biking when it rains. By doing so, she extends her topical orientation to enjoyable, embodied biking to also include biking when it rains.

Tine's turn proceeds throughout line 147 and onwards. In lines 147–8 and 151–2 she delivers a more fine-grained report on the embodied experience of biking when it rains, and especially her description of how she gets her pulse going and gets wet and sweaty should be noted. Then, in line 152 she marks the beginning of a concluding assessment with the stressed exclamation 'ej'/'oh my', and the following positive assessment comes off as being a matter of feeling alive, thereby further underlining the *embodiment* of the experience. The high granularity in Tine's account of the experience as well as her focus on embodiment stresses her orientation towards producing an *individualised, positive assessment* of biking when it rains; that is, towards accomplishing that she *herself* provides the fully competent reasons for her actions. Recalling Heritage, this highly specified report on a personal experience imposes a specific obligation on the recipient to exhibit empathic union with the describer. As I show below, this obligation is not fulfilled.

The moderator's response to Tine's orientation towards rain finally brings the analysis to the moderator's third attempt to focus specifically on riding a bicycle when it rains. As anticipated in the beginning of the analysis, the response comes off as a sort of 'third turn' following the second pair parts. In the aim to provide an understanding of the relevant sequential 'struggle' of the excerpt, some of the main features of such third turns should be considered. As Heritage (1984b: 257) mentions, 'third turn options' are reserved for the producer of the first pair part, and they provide the speaker with the opportunity of displaying his or her reception of the response to the turn. More specifically, in third turns, the speaker of the first pair part can deliver a number of different types of receipt tokens, each type performing a different operation on the second turn. For instance, the speaker can show that he or she is surprised at the second speaker's response, or that he or she would like to hear more, or that enough information has been given. The tokens can, for instance, be acknowledgement tokens, (partial) repeats, assessments, 'oh' receipts (which Heritage [1984a] characterises as 'change of state' tokens) and formulations, and, importantly, they can do different interactional work.

In the present excerpt, the formulation type is of specific relevance. Formulations can be characterised as a way of displaying a receipt of a second pair part through a formulation of a co-participant's (or, in this case, co-participants') talk. Garfinkel and Sacks (1970) describe a formulation in the following way: "A member may treat some part of the conversation as an occasion to describe that conversation, to explain it, or characterize it, or explicate, or translate, or summarize, or furnish the gist of it, or take note from its accordance with rules, or remark on its departure from rules" (350). Importantly, for present purposes, Heritage and Watson (1979) point out that formulations may play a role in the topical organisation of the conversation. They show that a speaker's formulation of the gist of the prior talk may function as the result of co-participants' negotiation about what that gist is. They further note that when co-participants accept such an agreed orientation, the formulation initiates the completion of the current topic. Thus, formulations in such cases come off as closure implicative. However, in the case of the moderator's third attempt, her formulation of the gist of the preceding, the participants do not straightforwardly accept the suggested agreed orientation displayed in it. Rather, in the remaining part of the interaction, the participants importantly negotiate the accomplishment of the gist; that is, of the topic.

In more detail, as the moderator in lines 155 and 157 says, 'så du tog cyklen til nakskov også når det regnede om morgenen'/'so you took the bike to nakskov also when it was raining in the morning', it comes off as displaying a receipt of Tine's second pair part work; that is, as displaying what the moderator understands to be the gist of the prior talk. Before looking into the more subtle interactional implications related to formulations of gist, it can straightforwardly be observed that not only does the moderator deliver her formulation at a point when an empathic response to a personal experience is due; she is also proposing an understanding concerning biking *to* work, not from work. Bearing this in mind, it is then useful to look into Heritage and Watson's work on formulations which very well observes the subtle interactional implications of such actions.

Heritage and Watson (1979) observe the interactional importance of formulations of gist, as they point to how formulations are drawn on in the accomplishment of conversations as self-explicating colloquies: "[Formulations] constitute members' method for providing that the conversation *has been* and is *ongoingly* a self-explicating colloquy: a specific instanced sense of the colloquy being retrievable *as part and parcel of this demonstration*" (139, emphasis in original). As this indicates, formulations constitute a powerful interactional tool, as they impose a membership-implicating challenge on their receivers. More specifically, Heritage and Watson argue that formulations are observably treated as first pair parts in the adjacency pair-format called 'formulation decision' (142). Furthermore, they suggest that formulations can be said to present a specifically constraining case. That is, as formulations involve the provision of candidate readings "and thus involve some demonstration or marking of [the sense established in preceding stretches of talk] in an instanced way, we may warrantably assert that they are deeply implicative for subsequent talk" (141–2). Considering this, it may not be surprising that there seems to be a preference for *confirmations* and a marked dispreference for *disconfirmations* as the decisive second pair part to formulations. Indeed, as already argued, the receiver of a formulation faces a highly membership-challenging task (144). Furthermore, Heritage and Watson draw on Pollner in their description of the interactional problem of dispreferred disconfirmations as a problem of "conversational 'reality disjuncture'" (146). Thus, a disconfirming second pair part plainly suggests that the parties to a conversation are not in the business of accomplishing a common world, but rather accomplish different worlds co-constituted by different memberships. Considering this, it may not be surprising that disconfirmations are rarely delivered without some sort of mitigation, some sort of display of their specific accountability.

However, it is important to acknowledge that formulations are not merely potential trouble initiators. Rather, as Heritage and Watson discuss in great detail, formulations are particularly useful or economical in topical talk. For instance, they note that when co-participants accept an agreed gist, the formulation smoothly initiates the completion of the current topic. Thus, formulations in such cases come off as particularly economical, closure-implicative third position turns. For present purposes, it is also useful to consider that Heritage and Watson observe that formulations of gist seem to have a particular role to play with respect to topical talk where gist has been the object of contest or negotiation. They observe that in such cases, formulations can be used to "establish the outcomes of such contests or negotiations 'for another first time'" (150). Furthermore, they point out that this use seems to be important in institutional interaction, where formulations are used to sum up meetings and where it is part and parcel of, for instance, 'doing being chairperson' to have the right to formulate (150).

Following these observations, when the moderator delivers a formulation in lines 155 and 157, it comes off as a typical part of 'doing being moderator'. Nevertheless, even though the formulation is a typical feature, it does not mean that the action does not have other ordinarily observed formulation implications. As it is, the fact that the formulation is implemented by a yes/no type of question, namely a declarative question, might even emphasise its moral implications. In other words, the formulation comes off as a somewhat imposing attempt to reattend to the issue of riding a bicycle no matter what, also if it rains, and, importantly, to mark this as the gist of the ongoing interaction (which of course includes the preceding). Hence, the formulation suggests that both the preceding and the ongoing interaction are concerned with accomplishing if not a 'no other alternatives but riding my bicycle' predicate, then an 'always riding my bicycle' predicate.

Importantly, Tine disconfirms the understanding displayed in the formulation, and as she does so, she also resists the predicate it invokes. However, the disconfirmation is mitigated as expected. As I demonstrate, the disconfirmation does not come off through a blatant rejection, but rather through a subtle negotiation of the predicate. Tine initiates her response to the moderator's formulation in line 158, and, in fact, the turn she delivers in that line comes off as a partly preferred confirmation insofar as Tine, without delay and straightforwardly, agrees saying, 'det gjorde jeg'/'i did'. However, the formulation is accomplished as a yes/no type declarative question. By implication, the *type* conforming preferred answer would have been 'yes'. Whether or not Tine's type non-conforming answer can be seen as being subtly aimed at negotiating the presuppositions in the formulation, it is followed by a short pause in which the moderator nods and, as it is, the sequence could have been over without further ado. This is not what happens. Rather, Tine proceeds in line 161, and as she does, her confirmation delicately turns into a carefully delivered negotiating disconfirmation. More specifically, in line 161 Tine starts mitigating the scope of the implied predicate by resisting its 'always-ness'. The mitigation comes off through the denying adverb 'ej'/'no' and the typical hesitation marker 'altså'/'well'. Furthermore, it should be noted that the mitigation comes off as commonplace through the commonplace displaying 'selvfølgelig'/'of course'. In this way, Tine subtly gets around the difficult membership-implicating challenges of disconfirming a formulation. She simply presents it as if it is already shared knowledge that the predicate was never supposed to be binding every single day.

In the following lines, Tine elaborates and accounts for the less binding predicate. She does so 'negatively', as she draws on a category-bound obligation to attend meetings. In lines 164 to 168 Tine uses the antecedent 'hvis jeg skulle til møde og alle mulige'/'if i was going to a meeting and everything' to explain or justify the consequent 'ah så tar jeg bilen i dag'/'ah then i'll take the car today'. The conditional design implies that if this obligation occurs, then there is a natural consequence. This necessity invokes a hearing of the obligation as not just any obligation, but as a *bound* and *binding* obligation, and one need only recall Sacks's economy rule to more specifically hear the obligation, that is, the predicate 'to attend meetings', as being bound to Tine's already accomplished 'employee' category. Consequently, the 'employee' category is accomplished as one of higher categorial relevance than any possible category with a 'riding the bicycle' obligation. It should be noted that the predicate comes off as particularly morally obligating due to the shift to present tense and reported thought in the consequence. Recalling Heritage's observations on reports on experiences, it can be argued that by drawing on such a fine-grained report, Tine's account comes off as more imposing.

By reformulating or respecifying the 'riding a bicycle when it rains' predicate invoked in the moderator's formulation, Tine 'sums up' or underlines the categorial settlement in lines 168 and 169. In lines 168 and 169, Tine subscribes to a less binding 'have biked, *also* when it rains' predicate. Then, in line 171, she claims ownership of the reasons for her subscription, providing the assessment that she sometimes enjoyed this practice of riding her bicycle in the rain. Remarkably, however, Tine clarifies the assessment by providing a more detailed description of the situations in which riding a bicycle in the rain is enjoyable. More specifically, Tine makes a distinction between a type of rain in which it is not enjoyable to ride a bicycle and a type of rain in which it is enjoyable. Thereby, Tine addresses the moderator's faulty understanding of what Tine's talk might imply, caused by the fact that the moderator failed to understand that Tine does not enjoy *all* kinds of rain. In this way, Tine of course underlines that you should not ride a bicycle when it rains, except if you enjoy it.

Lastly, as Tine provides her description of non-enjoyable rain in line 177, she uses the almost onomatopoetic verbal expression 'sjyuw', accompanied by a significant bodily expression. Again recalling Heritage, the moderator as well as Julie might be expected to immediately align and affiliate with this highly intense expression. Importantly, as the moderator does so, she simultaneously accomplishes Tine's displayed understanding that the moderator's formulation was formulated on the basis of a faulty understanding. Hence, in the end, the moderator fulfils Tine's accomplishment of a topic *not* on 'always riding a bicycle, also when it rains', but on 'enjoyable, fulfilling biking, sometimes also when it rains', and, by implication, Tine is able to resist the relevance of a 'no other alternatives' predicate. Thus, the analysis has shown how the topic '*always* riding a bicycle, also when it rains' is rendered irrelevant and how '*enjoyable* biking, *sometimes* also when it rains' is rendered highly relevant, thereby accomplishing biking as an enjoyable, fulfilling choice, once in a while, and not as a relevant alternative to automobility. This morally infused negotiation of *proper* conduct of biking conduct can be understood as a particular and skilled way of maintaining the governance of automobility at the molecular level.

The inevitability of cars

This section concerns how the car is approached as being an object that is always already possessed, thereby invoking a 'car owner' category that features certain indisputable account resources. The excerpts discussed in the following again stem from a session taking place before the municipal theme month, concerning the participants' everyday transportation practices. The moderator asked questions about walking, biking, going by car and using public transportation. The interaction represented in Excerpt Two/A begins approximately thirty-four minutes into the session. At that point, the parties to the setting have finished talking about walking and biking practices, and the represented interaction displays the first proffer and the first response to talk on practices of going by car. The interaction represented in Excerpt Two/B took place ten minutes later, around forty-five minutes into the session. At that point, the moderator proffers a slight topic transition, asking the participants what sorts of cars they drive (B, line one).

Excerpt Two A + B
Participants: Moderator: mod. Julie: jul. Sussi: sus. Berit: ber. Tine: tin. Also present: Pernille

A: 0.34.01

1	mod	.hh ja (.) men- (.) bilen- (.) .hh yes (.) but- (.) the car- (.)
2	jul	°ha-ha°
3	mod	så (.) hvor- (.) hvem hvem kører i den hvornår so where- (.) who who drives it when
4		(.)
5	jul	a'så f- (.) ligenu arbejder peter min mand jo i nakskov well f- (.) right now does peter my husband work in nakskov
6		så den bruger han jo hver dag .hh men når vi skal til you know so he uses it of course every day .hh but we also go

7 dans å så'n nogle ting så kører vi osse i bil
by car when we are going for dance classes and things like that

8 a'så (.) jeg havde da ikk' noget i mod å cykle men så skal
really (.) i wouldn't mind biking right but then you would

9 man jo ha barnepige i mange timer jo ha-ha hvis
of course also need a sitter for many hours right ha-ha if

10 vi skal cykle til nakskov [først=
we had to bike to nakskov [before=

B: 0.43.37

1 mod å (2.2) hva er det for nogle biler i kører i
and (2.2) what sorts of cars are you driving
((looks at jul))

2 (2)

3 jul vi har så'n en syvpersoners fordi vi::
we have like a seven persons because we::

4 mod °en=°
°a=°

5 jul =har så mange børn ha-ha-ha-ha
=have so many children ha-ha-ha-ha

6 mod hva øh hva øh ved du hva det er for et
what um what um do you know what brand

7 mærkes
it is

8 jul nej jo det en obel [ha-ha
no yes it's a obel [ha-ha

9 mod [ha-ha

10 jul za(h)fi(h)ra ha-ha
za(h)fi(h)ra ha-ha

11 [hi-hi-hi

12 mod [hvis du havde- hvis det nu
[if you had- if it was

13 havde været min mor så: (.) kun'
my mon then (.) it could

14 det faktisk godt ha' været
actually have been that

15 [hun ikk' hav(h)de vidst(h) hva
[she didn't know(h) what

16	jul	[ha-ha-ha.ha
17	mod	[de(h)t var for et mær(h)ke [kin(h)d of bra(h)nd
18	jul	[ha-ha-ha
19	ber	en rød bil a red car
20	All	[ha-ha-ha
21	mod	en obel zafira det så noget da- hvor an obel zafira it's like- where
22		der ka' klappes (.) [op it's possible to fold (.) [up
23	jul	[der [it's
24		ka' være syv sæder jaer de to possible to have seven seats yes the two
25		bageste ka ligges ned rearmost can be folded in
26	mod	jaer yes
27	jul	så: der jo kun tre (.) når man har then it's of course only three (.) when you
28		lagt dem ned så: der tre bagved fold it in then there are three at the back
29		å to ( ) and two ( )
30	mod	ja yes
31		(.)
32	jul	jaer yes
33	mod	[benzin [petrol
34	jul	[men nu] har vi så'n set ikk' så [but now] we do not really
35		meget brug for den mere fordi: (.) vi har jo need it as much any longer because (.) you know we
36		ikk' alle de børn mere den don't have all the children anymore the

37 sto(h)re han er jo ikk' hjemm(h)e mere
old(h)est you know he is not ho(h)me any longer

38 mod nej [men det kommer han vel så=
no [but he's at home i guess=

39 jul [den kører på benzin
[it runs on petrol

40 mod =alligevel i ny å næ
=anyway once in a while

41 jul jaer
yes

42 mod eller hvad
or

43 jul jaer å så ka peter bruge den han er
yes and then peter can use it he's

44 fodboldtræner så ka han ha nogen børn
a football coach so he can drive some of the chilren

45 med å så'n noget
and things like that

46 mod årh smart
oh-oh great

47 jul ja så det blir den osse brugt til
yes so it's also used for that

48 mod ja
yes

49 jul hi-hi-hi

50 mod å det benzin
and it's petrol

51 jul ja (.) å den kører vel omkring: (.)
yes (.) and it runs i guess around (.)

52 tolv tretten på literen
twelve to thirteen on a litre

53 mod ja
yes

54 jul (hvis du vil vide det osse) hi-hi-hi
(if you want to know that as well) hi-hi-hi

55 mod ja men det:: (.) a'så å (.) tænk-
yes but it's (.) well and (.) do you th-

56 tænker i (.) hvor tanker i henne (.)
do you consider (.) where do you refuel (.)

57 tanker i her i horslunde eller
do you refuel here in horslunde or

58 jul ja vi tanker tit i brugsen men det
yes we often refuel at the supermarket but it's

59 jo tem'lig dyrt
of course quite expensive

60 mod ja (.) så i- så i tænker ved hvis i
yes (.) so you- so you consider it if you

61 kommer forbi et andet sted
get by another place

62 jul å [det er billigere
and [it's cheaper

63 mod [å i er så heldige
[and you're so lucky

64 at det passer å så videre så
that it's on that time and so on then

65 jul ja
yes

66 (.)

67 jul [ja det gør vi
[yes we do

68 mod [så fylder i på der
[then you'd refuel there

69 jul det kigger vi altid efter når vi kører
we always look for that when we're driving

70 mod ja
yes

71 (.)

72 mod så i- i- i- i bevidste om at det tar
so you're concerned about that it takes

73 nogen penge
some money

74 jul helt sikkert ja
certainly yes

75 mod [ja
[yes

76 jul [ja] det tar da mange penge så'n
[yes] it takes really a lot of money things

77		[noget [like
78	sus	[um [um
79	jul	benzin fuel
80	mod	ja yes
81		(.)
82	mod	øhm: (1.1) f- (.) når man (1) bor her um (1.1) d- (.) when you (1) live here
83		å arbejder i nakskov så får man (.) and work in nakskov then you're (.)
84		lige præcis <u>ikk'</u> befordringsfradrag exactly not entitled to car allowance
85		eller hvad or are you ((looks at jul))
86	jul	nej det [tror jeg ikk' no i [don't think so
87	ber	[nej det passer vist [no that's right i think
88		[er det ikk' femogtyve kilometer [isn't it twentyfive kilometres
89	All	[( )
90	sus	de:t tolv kilometer its twelve kilometres
91	tin	der ska være meget længere it has to be much further
92	mod	ja yes
93	sus	du ska køre fireogtyve kilometer you'll have to drive twentyfour kilometres
94		om dagen a day
95	ber	ja yes
96	mod	ja å det nemlig bare om dagen så yes it's right it's only a day so

97		man ku sige det sku bare lige være you could say it just had to be
98		lidt længere= a little further
99	sus	ja yes
100	mod	=end nakskov så (.) var der et par =than nakskov then (.) it would release
101		kroner å hente men .hh så det er- a few krones but .hh so it's-
102		det er af egen lomme it's out of your own pocket
103	jul	jaer så han burde cykle yes so he should be biking
104	mod	ja yes
105	jul	men de(h)t ha-ha får jeg but th(h)at's ha-ha
106		ham ikk' til ha-ha not gonna happen ha-ha
107	mod	ha-ha har du foreslået det ha-ha have you suggested that
108	jul	nej han har ikk' engang rigtig nogen no he doesn't even really have a
109		cykel bike
110	mod	nej (.) så du har ikk' al'så du har ikk' no (.) so you haven't well you haven't
111		engang så'n antydet det så'n even like indicated it like
112	jul	nej det no i
113		[ka je lige så godt lade være med [could just as well leave it
114	mod	[indirekte [indirectly
115	jul	jeg ved det ikk' blir til noget ha-ha i know it wouldn't be going to happen ha-ha

116 mod ja
yes

117 (.)

118 mod å: men ø:h å >hva med bussen<
and but um and what about the bus

119 (1.2)

120 jul a':s' de:t da lige så dyr som benzin næsten=
well it's almost as expensive as fuel=

121 mod [ja
[yes

122 jul [=det] koster da tredve kroner hver vej
[=it's] surely thirty krones each way

123 (sus) °uhm°
um

124 (.)

125 jul så: det [vil' han jo ikk' gøre
so of course [he wouldn't do that

126 mod [men det ville det
[but then it

127 jo ikk' gøre hvis man <gjorde det> hver dag
wouldn't if you did it every day

128 jul så:'n et værdikort nej det vi' nok blive lidt
like a discount card no it'd probably be a little

129 [billigere
[cheaper

130 ber [nej] det blir billigere
[no] it'd be cheaper

131 jul [ja
[yes

132 mod [ja
[yes

133 jul .hh men øhm=
.hh but um=
((crosses arms))

134 mod =a'så jeg- jeg [ved=
well i- i [of course

135 jul [ja
[yes

136 mod =jo bare hvad- mellem (.) prisen mellem
= i only know what- between (.) the cost between

137 aalborg å odense om jeg så må sige men det-
aalborg and odense so to speak but it-

138 (.) det a'så .hh hvis man (.)
(.) well it .hh if you (.)

139 [kører ofte så-
[go often then-

140 jul [det rigtigt det falder (jo) efter
[that's right it's cheaper (of course) the

141 flere gange ja
more times yes

142 mod så- så går det faktisk hen å blir (.6)
so- so actually it ends up being (.6)

143 [decideret
[kind of

144 jul [ja
[yes

145 mod billigt (.8) det er- de:t stadig billigt å være
cheap (.8) it's- it's still cheap to be

146 pendler
a commuter

147 jul ja
yes

148 mod men det er godt nok ikk' billigt å ta enkeltture
but it's definitely not cheap riding single rides

149 jul nej (.) .hh >jamen nu har vi jo< bilen a'så
no (.) .hh but now we do you know have the car really

150 mod [ja
[yes

151 jul [når] jeg ikk' bruger den så ville det jo
[since] i don't use it then it would you know

152 mod ja
yes

153 jul da nok
surely i guess

154 [være dumt at han tog bussen ikk' (a's')
[be stupid if he took the bus right (really)

155 mod [al'så men nå men- men- men- men a'så men
[well but then but- but- but- but well but

156 (.5) det var- det var osse mere om det var- (.7)
(.5) i was- i was rather thinking like if it was(.7)

157 n- noget i havde- (.5) ((gestikulerer))
s- something you had (.5) ((gestures))

158 mod [det ku godt være i havde haft det oppe å vende
[it could be that you had been considering it

159 jul [vi har haft snakket om på et tidspunkt om vi overhovedet
[we did talk about it once whether we

160 behøvede bilen nu hvor det var men
needed the car at all now as it is but

161 mod ja
yes

162 jul a'så det jo meget rart å ha osse når man skal på
well it's you know pretty nice to have also when you're going on

163 ferie å når vi ska: ud å køre å:
holiday and when we're going for a drive and

164 (.) [på=
(.) [to the

165 mod [ja
[yes

166 jul bunkeren å så'n nogle steder [ellers=
tip and places like that [otherwise=

167 mod [ja
[yes

168 jul =ka man jo slet ikk' komme nogen steder jo
=you cannot really go anywhere you know

169 mod nej
no

170 jul så:
so

171 mod nej å i: mange (.)
no and you're are quite a few (.)

172 [i: s-
[you're s-

173 jul [ja
[yes

174 mod eller i: stadigvæk (.) i hvert fald
or you're still (.) after all

175	jul	ja (.7) med [tre børn= yes (.) with [three kids
176	mod	[nogen stykker [quite a few
177	jul	=det jo osse stadigvæk =well you know it's still
178	jul	ja yes
179		(.)
180	mod	((looking at ber)) hva- hva med dig hva er det- hva kører du i what- what about you what is it- what are you driving
181	ber	øh en obel korsa um a obel korsa

The focus of the analysis is the interaction in Excerpt B and, more specifically, the interaction that begins with (what is interactionally accomplished as) the moderator's information-eliciting question in line 118. In the interaction following this question, the category 'car owner' is drawn on as an account resource. Furthermore, 'car-mobility' is accomplished as the in situ relevant form of mobility, in that 'to go by car' is settled as *the* modus operandi concerning both everyday and holiday mobility.

However, while I am particularly interested in the interaction represented in lines 118 to 178, it is necessary to consider a much broader part of the interactionally accomplished context to be able to comprehend the subtle references in the lines in focus. Hence, I would like to begin by providing a rough overview not only of the interaction displayed in Excerpt B, but also of how this interaction is related to the interaction represented in Excerpt A. Excerpt A displays the first proffer and the first response concerning going by car practices. In lines one and three, the moderator asks, roughly, 'What about the car, who drives it and when?', and after a short pause, Julie self-selects in line five and fulfils the moderator's proffer, providing a rich second pair part to it. In the nine minutes of not-displayed interaction, Julie finishes her second pair part, and the parties to the setting accomplish a round of second pair parts to the proffer. In these nine minutes, nothing happens which is of particular interest to the present analysis. Of course, a lot happens in nine minutes, and the fact that I have omitted the interaction is primarily due to the concerns of providing a reasonably focused analysis.

The interaction displayed in Excerpt A is relevant to my analysis of Excerpt B in two ways. The first reason is more sequential, whereas the second is, to a greater extent, related to the relevant interactional knowledge accomplished in this interaction. Firstly, line one in Excerpt B comes off as projecting a topic transition on the backdrop of the preceding accomplishment of the going by car topic. Recalling Svennevig (2000), a topic transition is a minor adjustment to an already established topic (210).[8] Secondly, in lines five to ten, Julie provides pieces of information that are referred to as common knowledge in Excerpt B. Particularly, Julie reveals that since her husband, Peter, presently works in Nakskov, he uses their car to go to work every day (lines five to six). Furthermore, she adds that they use the car when they go to dance classes and 'things like that' (line seven), and she adds that she would not mind biking, but that they would need a sitter that could stay for many hours in a row if they had to ride their bicycles to Nakskov (lines eight to ten).

Considering Excerpt B, overall the excerpt displays the moderator's proffer for a topic transition and Julie's delivery of a first second pair part in what seems to progress as yet another multiple second sequence. The proffer for a topic transition is projected in line one when the moderator asks, 'og hvad er det for nogle biler i kører i?'/'and what sort of cars are you driving?'. The moderator looks at Julie, and after a pause, Julie fulfils both the next speaker invitation and the topic transition. In other words, in line three, Julie initiates a second pair part, thereby initiating the topic *talking* sequence itself; that is, the specific form of substantial expansion generated by topic encouraging seconds (cf. Schegloff, 2007a: 180). Furthermore, in line 180 the moderator is once again 'doing being moderator' on the surface as she unilaterally finishes the expansion; that is, the topic talk generated by Julie's second pair part. In line 180, she turns to Berit and asks her what kind of a car she drives, and in this way, she reveals that the parties to the setting are doing a round, drawing on the format of a multiple seconds sequence.[9]

---

8 Concerning the claim that the moderator orients to line one in Excerpt A and line one in Excerpt B as projecting a new topic and a topic transition, respectively, the difference in orientation can be noted even at the most superficial level. In Excerpt A, the moderator takes a breath, says 'yes' and then 'but the car'. While the breath and the 'yes' can be heard as marking a transition, the transition is more readily heard in the disjunctive 'but', which marks a difference to the preceding, and in the definite article (in Danish the suffix '-en', in English 'the'), which displays a generic orientation to the car as a topic in its own right. On the other hand, line one in Excerpt B is initiated with a tying 'og'/'and', which ties it to the preceding.

9 Schegloff (2007a) observes that due to such preferred topic talking expansions, ending such sequences can lead to interactional trouble (180). However, in extended sequences, one way to do sequence closing through 'unilateral' sequence ending is having one of the parties to the interaction quite simply take on the power to proceed. It is part and

However, within the accomplished boundaries of this overall structure, lines eighty-two to 178 seem to display the accomplishment of some sort of side sequence (Jefferson, 1972). As demonstrated in more detail below, the interaction in these lines seems to be oriented to the accomplishment of background information that is not relevant in its own right but supports or elaborates or problematizes what has been established in Excerpt A.[10] Hence, my main focus is embedded in this side sequence. Considering this, the rougher outline of the excerpt is not enough to provide a proper starting point for an analysis of line 118 and onwards; it is, in fact, necessary to consider the entire side sequence in which it is embedded. Consequently, I begin by providing a more superficial introduction to the first half of the side sequence, from lines eighty-two to 116. Against this, I secondly turn to a more thorough analysis from line 118 onwards.

As stated, the moderator's utterance in line eighty-two[11] comes off on the backdrop of the topic-talk generated by the moderator's proffer in line one and Julie's response in line three. Hence, lines one to eighty-one should briefly be considered, focusing only on the elements of relevance to my enquiry into lines eighty-two to 116. Whereas the pause in line two is notably long, no other features mark Julie's response as dispreferred. Both verbally and visually, the pause comes off as an expectable matter of 'doing being parties to a focus group setting'. That is, the pause seems to reveal that Julie is simply making sure that the moderator has finished, and that her gaze is in fact selecting Julie as the proper next speaker. Accordingly, in the following topic talk, Julie provides a rich account of her and her husband's car, almost explicitly ascribing to her co-constituting memberships of 'proper speaker on the topic' and 'party to a focus group setting'. For instance, in line thirty-four, Julie overlaps a turn initiated by the moderator and delivers an unproffered elaboration, and in line fifty-four she makes her role

parcel of 'doing being moderator' that the moderator often closes the multiple second pair in exactly such a unilateral way (181–6). Further, in relation to the phenomenon of 'doing a round', note Schegloff's notion of multiple second pair parts sequences; these are sequences in which multiple speakers respond to a first pair part (161, n 23).

10 Again, I draw on Svennevig (2000) for my understanding of the use of side sequences in topic talk. He observes, "The information supplied or elicited in these side sequences is marked as background information in that it is not presented as relevant in its own right, but only as support for some presentation in the previous or subsequent discourse. The talk about the background information is abandoned once it has supplied the necessary background for continuing the main topic. What happens in such sequences is that the background is foregrounded or made explicit in a special way" (257).

11 From this point, all line numbers refer to Excerpt B, unless otherwise stated.

as participant observably relevant as she comments on a piece of information she has just delivered, saying 'hvis du vil vide det også'/'if you want to know that as well'.[12] Hence, it should be feasible to hear the interaction as an unproblematic instance of 'doing being party to a focus group setting'.

Briefly considering the knowledge accomplished in those lines, Julie once again refers to her husband's use of the car. After Julie's initial talk about the sort of car they drive, she mentions that they really do not need this specific, rather spacious car now, since they no longer have many children living at home (lines thirty-four to thirty-seven). The moderator suggests that sometimes they may still have to fit the entire family into the car (lines thirty-eight and forty). Then, in lines forty-three and forty-four (and forty-one) Julie agrees and says, 'og så kan peter bruge den han er fodboldtræner'/'and then peter can use it he's a football coach'. The 'og så'/'well then' ties the incident to the preceding, and in this way, Julie co-accomplishes both the moderator's observation that they may still have to fit many people into the car and her own observation that her husband is a football coach, thus *accounting* for the fact that they *should* keep this particular car. Importantly, Julie's reference to her husband reflects the new topical environment. In Excerpt A, the proffer concerned the car and driving in general, and at that point, Julie drew on her husband's daily work commute to Nakskov as a resource that sustained the topic. Here, on the other hand, the proffer concerns the sorts of cars the participants are driving, and Julie draws on her husband's membership of 'football coaches' as a resource for accounting for the type of car they drive. There is no other reference to Julie's husband in the interaction between line one in Excerpt A and line eighty-two in Excerpt B, and after this football coach account, the talk moves on to petrol; more specifically, to how many kilometres Julie and her husband's car can go on one litre (lines fifty-one and fifty-two), and then on to fair prices of petrol (lines seventy-two to seventy-four and seventy-six to seventy-eight).

Against this, it is time to look into the turn initiated by the moderator in line eighty-two. The turn comes off as a negatively formatted yes/no type of question, requesting conformation concerning whether the moderator is right in believing that the distance between Horslunde and Nakskov is *not* great enough to yield car allowance. Lines eighty-two to eighty-four are formatted as a declarative, but

12 As noted in the transcript, the line is not quite audible. It is, however, fairly qualified to say that Julie is asking whether the information delivered is relevant to the moderator. Her laughter in the end of the line supports the understanding that she is concerned not only with 'doing being party to a focus group setting', but with doing *well* in her part of the accomplishment.

this is turned into a formal question by the tag question in line eighty-five.[13] Accordingly, whereas the question makes a 'no' the type-conforming preferred confirmation, it should, as always, be considered whether or not the question could be heard as tilted towards a type-conforming, but disconfirming positive 'jo' (yes2).[14] However, it cannot be considered before examining in more detail the moderator's work in these lines and the participants' responses to it.

Hence, firstly, I will comment on some of the moderator's exact wording. In lines eighty-two to eighty-three, she says, 'når man bor her og arbejder i nakskov'/'when you live here and work in nakskov'. The Danish 'man' is different from the English 'you', in that it could not be confused with an ordinary second person reference. At the same time, however, it is much less formal than the English 'one'; it is widely used in everyday interaction, most often to display a sense of a commonplace moral order. Yet, whereas the 'man' might display a certain 'everybody-ness', the moderator illuminates her request with reference to a situation notably similar to that of Julie's husband. Nobody but Julie has mentioned anyone who lives in Horslunde and works in Nakskov. Furthermore, the moderator delivers her request in a part of the interaction in which Julie is 'doing being the designated second pair part speaker', and, similarly, the moderator looks at Julie. Hence, even though the use of the pronoun 'man' reveals that the moderator is using a random example, the moderator is unambiguously oriented to Julie as the proper respondent.

Furthermore, in line eighty-six, Julie confirms the negatively formatted request through an undelayed and type-conforming 'nej'/'no'. Moreover, in the following lines other participants join in, and it is commonly established that it is correct that the distance between Horslunde and Nakskov is precisely too short to yield car allowance. Accordingly, there is apparently no orientation towards the moderator's request as being tilted towards a disconfirming 'jo' (yes2) or as being specifically morally implicating. In lines ninety-six to 102, the moderator sums up, concluding in lines 101 to 102, 'så det er af egen lomme'/'so it's out of your own pocket'. Then, in line 103, Julie says, 'ja så han burde cykle'/'yes so he should be biking'.

---

13 Heinemann (2010) observes that there are three different tag question formats that can be used to turn a declarative into a formal question in Danish. In addition to the so-called *wh*-type (in Danish the 'hv-type') used in this example, she mentions sentential tags and adverbial tags (2707).

14 Again, recall Heinemann's (2005) observation (which speakers of Danish know well) that a confirmation of a negative statement takes a negative form (versions of 'nej'/'no'), but that negatively formatted utterances *can* be designed to go against this general preference and to aim to disconfirm the literal assertion through a 'jo' (yes2) (393–4).

As it is, Julie's turn accomplishes a number of interrelated things. Firstly, considering the established interactional context, the personal pronoun 'han'/'he' comes off as an unambiguous reference to Julie's husband Peter, and 'he should be biking' comes off as referring to Peter's known work commute. Secondly, the pointedly *anaphoric* personal pronoun displays an understanding of the preceding interaction as *already* concerning Peter, and, consequently, Julie's choice of 'han'/'he' de facto accomplishes the moderator's question about car allowance as referring back to the conversation ten minutes earlier when the car topic was first proffered; that is, to Julie's second pair part in lines five to ten in Excerpt A. Hence, whereas the moderator's car allowance question might be heard as being linked to and elaborating on the preceding petrol talk, Julie's response in line 103 instead accomplishes it as initiating a side sequence oriented to the accomplishment of background information that might *problematize* what was established in Excerpt A, namely that Peter drives their car to work in Nakskov. Thirdly, as just indicated, although the background information could be treated as simply supporting an earlier accomplishment, Julie reveals that she approaches it as 'doing problematizing' through her use of the morally implying modal verb 'burde'/'should' (or, perhaps, 'ought to'). That is, Julie demonstrates that since it does not yield car allowance to commute between Horslunde and Nakskov, her husband is to some extent accountable for doing it.

Here it is time to revisit the question of whether the moderator's question makes 'no' the type-conforming preferred confirmation, or whether it could be heard as being tilted towards a type-conforming, but disconfirming positive 'jo' (yes2). Considering, firstly, that Julie displays an understanding of the interaction as a matter of accomplishing background information for her much earlier account and, secondly, that this information problematizes this account, she displays an orientation to the moderator's account as being tilted towards a positive 'jo' as the *morally* preferred response, in that she displays an understanding that implies that if her husband's commuting had yielded car allowance, he would have been less accountable for going by car.

Concerning this first part of the side sequence, the last, but important, thing to observe is that the moderator embraces and supports Julie's understanding that they are in the business of accomplishing a side sequence. Whereas Julie goes on laughing and pointing out that she would never be able to convince her husband to go by bicycle, the moderator, who initially joins in the laughter, asks Julie whether she has in fact tried to convince him, directly (line 107) or indirectly (lines 110, 111 and 114). Julie, however, rejects the possibility that she could be accountable for asking him, referring, firstly, to the fact that her husband does not even 'rigtig'/'really' have a bike (lines 108 to 109) and, secondly, to the fact that she 'ved'/'knows' that it would never happen (line 115). Again she laughs at the end of the last epistemically grounded account. Hence, whereas Julie, in line 103, reveals that her understanding of the interaction is that they have accomplished a piece of background information which renders her husband somewhat accountable, she does not consider her own suggestion of a 'solution' a serious possibility, but only a laughable 'joke'. In other words, the basis for the moderator's utterance in line 118 seems to be that even though petrol is regrettably expensive and even though it is regrettable that the distance is too short for car allowance, there really is nothing to be done about any of it.

In line 118 the moderator asks, 'og men øh og hvad med bussen'/'and but um and what about the bus'. Before looking into how this utterance is fulfilled in Julie's response, I would like to provide a couple of observations about so-called *wh*-questions. Heinemann (2010) provides a useful overview of the question response system in Danish. On the subject of *wh*-questions, she observes, "The most common matter that wh-questions inquire about in Danish is Object (e.g. 'what is that'). This is done with the hv-word hvad" (2711). In expected accordance with this observation, Heinemann's study shows that *wh*-questions are very often used for the social act of requesting information (or initiating repair). Accordingly, it would be reasonable to hear the moderator's line as 'doing requesting information'.

Considering Julie's response in line 120, it is audibly delayed and initiated with the typical dispreference marker 'altså'/'well'. Furthermore, Julie's lines 120, 122 and 125 display the account that the bus is almost as expensive as petrol, and that it costs thirty kroner, 'så det ville han jo ikke gøre'/'so of course he wouldn't do that'. The account is marked not only through the initial display of dispreference, but also through the common conclusion marker 'so' which demonstrates that the last line necessarily follows from the observation on the price of a bus ticket. The self-evident rationality of the account is underlined by the presence of epistemic adverbs, both in what is treated as the 'antecedent' and what is treated as the 'consequent'. As shown by Heinemann (2009), the epistemic adverb 'da' in answer turns "serves to target and reject a position taken (implicitly) by the questioner" (174).[15] Following this observation, Julie's statement, 'det koster da tredve kroner'/'it's surely thirty kroner', appears to reject what is simultaneously accomplished as the position taken by the moderator, namely that Julie's husband could

15 Furthermore, Heinemann (2009) observes that the presence of 'da' can be used to display a prior action "as *morally* problematic", and as such it can serve "to rebuke the recipient for her failure to take shared knowledge or values into account" (178). See also Heinemann, 2008: 63.

(or should) go to work by bus. Furthermore, the self-evident impossibility of the moderator's suggestion is cemented through the self-evident epistemic adverb 'jo' in the conclusion (here translated into 'of course', but in other cases it could be 'you know').[16] In this way, Julie's response comes off as being unambiguously dispreferred and as treating the moderator's question as self-evidently implying something irrational.

Revisiting the moderator's *wh*-question, it can, accordingly, be seen that it is not an in situ 'innocent' request for information, but rather morally loaded. As hinted at, this pointedly moral accomplishment might be due to two inevitably related features. Firstly, it might be yet another expected instance of 'doing being parties to a focus group setting', and, secondly, it might be due to the actual sequential environment. In more detail, it should be recognised that the moderator's question comes on the backdrop of Julie's suggestion in line 103, namely that her husband should be biking. In other words, the moderator's question comes off as a 'next' in a list of possible alternatives to Peter's accomplishment of accountable driving.

Before leaving Julie's account, I would like to consider the category work accomplished in and through the simultaneous accomplishment and rejection of the moderator's suggestion. In more detail, Julie points out that going by bus is 'næsten'/'almost' as expensive as buying fuel (line 120); this is underlined by the specific price mentioned in line 122. Both lines display the price as *high* and the first, moreover, underlines this by comparing the prices of going by bus and going by car (indirectly done through the reference to fuel prices). Considering the sequentiality, it is thereby accomplished that it would be accountable to go by bus, when it is *almost* as expensive as going by car, or, in other words, Julie's husband cannot be held accountable for not going by bus; quite the opposite in fact, he has an *obligation* not to do so.

However, the moderator does not readily acknowledge Julie's account work. On the contrary, the moderator interrupts the account when she starts to talk in a *non*-transition relevant place in line 126, initiating her turn with the clearly

16 Heinemann (2008) suggests that the presence of this epistemic adverb marks what is said as self-evident (60). See also Heinemann, Lindström and Steensig, (2011: 107–30). They provide a thorough discussion of how 'jo' is used in different sorts of morally imbued knowledge negotiations. However, they also point out that while the adverb can be said to carry a generic function, namely some kind of 'appeal to shared knowledge', the more specific implications should be considered in every single case "through paying attention to the various sequential positions in which the adverbs are produced" (130).

disjunctive 'men'/'but'. Notably, the moderator does not display disagreement with the *logic* of Julie's account. The moderator does not negotiate whether it is accountable to go by bus if it is almost as expensive as going by car (for instance, the moderator does not ask whether it would not be better to save even a small amount of money by going by bus). Rather, in lines 126 to 148, the moderator negotiates whether it *is* in fact almost as expensive to go by bus, or whether it is, perhaps, cheaper after all.

Through this negotiation, what is accomplished is, firstly, that it is accepted that the bus has to be more than just a little cheaper than the car to render the bus a non-accountable alternative, and, relatedly, that the car is the standard when measuring transportation practices. Some of the finer details of this work should be noted. Firstly, whereas Julie agrees with the bottom line of the moderator's argument throughout these lines, her body displays withdrawal, as she crosses her arms in line 133. This might not be relevant, but the moderator can be said to treat it as such. Hence, secondly, in lines 134 and 136 to 137, the moderator predicates herself as someone that 'ved'/'knows' (lines 134 and 136) what the price is for a train ticket from Odense to Aalborg. The achievement of this category work comes off as two related aspects. The categorisation mitigates the moderator's interrupting negotiation of Julie's account, in that it limits the scope of the moderator's knowledge, underlining that the moderator 'bare'/'only' (line 136) knows what it costs to go by train between Aalborg and Odense. Since these cities are placed in other parts of the country than the one presently discussed, the categorisation devaluates the moderator's negotiation of Julie's account. Furthermore, since the point of the moderator's categorisation is the scope of her knowledge of *prices*, the categorisation emphasises the common orientation towards *expenses*.

Hence, the moderator, on a more overall level agrees with Julie's logic, and her geographical category work mitigates her objection to Julie's account. Lastly, however, it should be noted that the moderator continues in line 137 by accomplishing a more general 'pendler'/'a commuter' category (line 146), which is characterised by the existential predicate 'billigt og være'/'cheap to be' (line 145). Accordingly, in the end, lines 126 to 148 do display a heavy negotiation of the point of Julie's account, namely that it would be nearly as *expensive* to go by bus.

Thus, it is not surprising that Julie provides yet an account in lines 149 to 154. Julie initiates line 149 with a 'nej'/'no' that comes off as agreement with the moderator's line 148 where she says that one-way tickets are definitely not cheap. However, the rest of Julie's turn is clearly oriented to the moderator's overall negotiation of Julie's first account. After the 'no' and a brief pause Julie takes an audible breath and says, 'jamen nu har vi jo bilen altså'/'but now we do you know have the car really'. The 'jamen'/'but' displays a disjunctive contrast to the above which is more marked than it would have been had it just been 'men'/'but' (hence, the English version could perhaps also be 'well but'). Following, Julie refers to the fact that they do have the car, and her use of 'nu'/'now', 'jo'/'you know' and 'altså'/'really' demonstrates this state of affair, not only as *the* present state of affairs (invoked by 'nu'), but also as a state that *should* already be known (invoked by 'jo') and which invokes certain evident implications (invoked by 'altså' which is perhaps more explicitly causality-implicating than 'really'). However, the interactional role of 'jo' should be considered in more detail.

Recalling the 'jo' in Julie's first account (line 125), it refers back to the piece of information she has just provided concerning the price of a bus ticket (lines 120 and 122). Despite the fact that the price reference is delivered as rejecting some assumption in the moderator's suggestion, the 'jo' in line 125 comes off as a fairly affiliative 'jo', namely as a 'kind reminder' of the by now shared knowledge of bus prices (cf. Heinemann et al., 2011: 118–23). On the other hand, here in the first line of the second account, 'jo' seems to co-constitute a somewhat disaffiliating account, indexing some sort of moral transgression. Heinemann et al. (2011) observe that such 'jo' accounts "address the moral implications" of one or another epistemic incongruence, "laying the blame for this incongruence squarely at the feet of the questioner" (127). However, as they further observe, it depends on a closer analysis of the actual activity to decide whether the 'jo' in question is in fact "blaming the questioner" (127). Accordingly, it is necessary to look into the remaining part of Julie's account to determine whether her 'jo' in line 149 opposes the thereby co-accomplished moral stance in the moderator's suggestion "what about the bus".

In line 151, Julie says, 'når jeg ikke bruger den så ville det jo'/'since i don't use it then it would you know', and after the moderator's 'yes' in line 152, Julie continues her turn in lines 153 to 154, saying 'da nok være dumt at han tog bussen ikke (altså)'/'surely i guess be stupid if he took the bus right (really)'. These lines feature not only yet another 'jo', but also a 'da', which, it should be recalled, can be used to present a prior action as *morally* problematic. Additionally, Julie explicitly targets the moderator's suggestion and treats it as a moral transgression, saying that it would be 'dumt'/'stupid' (line 154) for her husband to take the bus. Lastly, if this is not enough to support the notion that Julie is holding the moderator morally accountable, the moderator's response should finally be noted.

The moderator initiates her response in line 154, and two things are right on the surface. Firstly, the first line displays an overwhelming degree of retreat. The hedging 'altså'/'well' and the repeated self-repair on 'men'/'but' come off as initiating the repair, and, importantly, this orientation is continued in the following lines (lines 156 to 158) in which the moderator suggests another hearing of her first request, namely that it 'kunne være'/'could be' that Julie and her husband 'havde haft det oppe at vende'/'had been considering it' (line 158). Secondly, the moderator's first line overlaps with Julie's line 154; that is, the moderator initiates the repair *before* Julie's argument that it would be stupid for her husband to go by bus. Consequently, even without Julie's explicit targeting, the moderator approaches Julie's account as *rightfully* holding the moderator morally accountable.

Significantly, the account is, of course, generated on the basis of the observable and indisputable relevance of Julie and her husband's membership of the 'car owner' category. The category itself is invoked through the naming of the relationship that binds it, namely Julie's statement that they do have the car (line 149), and its indisputable relevance is exactly part of the point of the account, made observable through the already mentioned features invoking that this categorial relationship is to be known as sedimented. That is, Julie's 'nu'/'now' characterises the category-invoking relationship as an inescapable *present* state of affair, while her 'jo'/'you know' presents this state as a 'should be known in common' state. Additionally, the point of the account is also that the car owner membership entails an obligation, rendering Julie's husband obligated to go by car. This is co-constituted by Julie's remarks that *she* does not use the car (line 151) and that it would be *stupid* for her husband to go by bus (line 154). Recalling the moderator's accepting, repairing response, it is accomplished as common knowledge, shared by all reasonable members, that if you own a car, you *have* to use it and, therefore, cannot go by bus. That is, notably, the moderator accepts Julie's view of the *present* world as not accountable.

Considering the last part of the excerpt, Julie accepts the repaired version of the moderator's request. In fact, Julie more or less co-formulates it; not only does her line 159 overlap with the moderator's line 158, it confirms it even as it is being produced. Whereas the moderator's reworked request comes off as a mere *suggestion* that it could have been the case that Julie and her husband had considered 'det'/'it', without explaining what 'it' is, Julie's overlapping response comes off as the perfectly preferred granting of the information that 'vi har haft snakket om på et tidspunkt om vi overhovedet behøvede bilen'/'we did talk about it once whether we needed the car at all' (lines 159 to 160). Thus, fulfilling the moderator's request, Julie can be heard as raising a possible issue concerning car ownership, namely that it might not, after all, be necessary to own a car and to insist on membership of the car owner category. However, Julie readily counters this possibility, initiating it in the same turn. Hence, Julie ends line 160 with a disjunctive 'men'/'but', thereby initiating her counter-argument, and, furthermore, after the moderator's acknowledging 'yes' in line 161, Julie proceeds in line 162 and onwards, countering the possibility that it might not be necessary to own a car, using both a convenience and a necessity argument. Consequently, the job of the counter is to ensure that the car owner membership and the subsequent obligation to go by car are, all things considered, a necessity.

The counter is initiated with Julie's point that it is 'meget rart'/'pretty nice' to have a car (line 162), and, by implication, the first reason for car ownership is the 'niceness' about it. Then Julie specifies in more detail how she understands this niceness (and, as it is, necessity). She lists a number of activities that in different ways illuminate her argument. She points out that it is nice 'også når man skal på ferie'/'also when you're going on holiday' (lines 162 to 163), and she proceeds, adding 'og når vi skal ud at køre og på bunkeren og sådan nogle steder'/'and when we're going for a drive and to the tip and places like that' (lines 163 to 164 and 166). The activities are displayed as commonplace activities through the certainty of the time conjunction 'når'/'when' (compared, for instance, with a hypothetical conjunction like 'hvis'/'if') and through the categorical verb modality 'skal'/'are going' (compared, for instance, with 'vil'/'would like to go'). Notably, in lines 166 to 168 Julie summarises this list, pointing out that 'ellers kan man jo slet ikke komme nogen steder jo'/'otherwise you cannot really go anywhere you know'. Consequently, Julie suggests a moral order in which 'to go by car' is established as *the* modus operandi concerning both everyday and holiday mobility.

Lastly, note that Julie's argument is delivered with no less than two commonplace-displaying 'jo'/'you know' utterances, namely in the beginning and in the end (lines 162 and 168). These adverbs strengthen the already morally loaded argument, and, as it is, the argument is not contested. Rather, the moderator accepts this moral order with a preferred confirming 'nej'/'no' in line 169, and in the remaining lines, she even adds to it, reiterating the point that Julie's family includes quite a number of people (lines 171, 174 and 176). Then, in line 180, the moderator ends both the side sequence and Julie's second pair generated topic talk by turning to the 'next' participant in the thus projected multi second pair part sequence.

Summing up, the analysis has shown how the category 'car owner' is invoked as an ultimate account category. It has shown that the category is accomplished as co-constituted by the category-bound obligation to go by car and that this obligation is used as a resource not only for accounting for why Julie's husband does not go by bus, but also for holding the moderator accountable for indicating that he might. Furthermore, the analysis has shown that 'car mobility' is accomplished as the in situ only relevant form of mobility, in that 'to go by car' is established as *the* modus operandi concerning both everyday and holiday mobility.

Summary

The above analyses both concern excerpts from focus group settings conducted prior to the municipal theme month. In these sessions, the moderator generated topic talk on everyday transportation practices, thereby making rationalities and moralities that co-constitute the governing of automobility available for enquiry in the face of a simple call to account for one's own practices. In other words, these sessions display the hows of the practices concerned to conduct one's own conduct.

The first analysis demonstrated how biking is subtly, but observably accomplished as a non-obligation in and through one of the participant's careful resistance to a 'no other alternatives' predicate. In more detail, it showed how the moderator invokes and draws on a 'no other alternatives' predicate, as she proffers a topic on 'always riding, also when it rains'. However, the responding participant, Tine, on the other hand, carefully accomplishes a topic not on 'always riding a bicycle, also when it rains', but on 'enjoyable, fulfilling biking, sometimes also when it rains'. As demonstrated, Tine approaches this accomplishment as doing the job of a completely preferred, topic-encouraging second pair part to the moderator's proffer, and, thereby, she resists the relevance of the 'no other alternatives' predicate. Hence, the analysis showed how 'always riding a bicycle, also when it rains' is not even considered a possible topic, and how 'enjoyable biking, sometimes also when it rains' is rendered highly relevant. In other words, the analysis showed how riding a bicycle was accomplished as an occasional, enjoyable and fulfilling choice, not as a relevant alternative to car driving.

The second analysis demonstrated how Julie accomplishes a 'car owner' category in a response that heavily negotiates the moral implications of a request from the moderator. The analysis demonstrated how Julie accomplishes the category as a forceful account resource which invokes certain notably significant obligations. Furthermore, it showed that 'car mobility' is accomplished as the in situ only relevant form of mobility, in that 'to go by car' is accomplished as the modus operandi concerning both everyday and holiday mobility.

I would like to suggest that the two analyses provide insights into the rationalities that co-constitute the governing of automobility. Particularly, I would like to suggest that both analyses could be said to be highlighting rationalities that co-constitute the maintenance of automobility. Whereas this rationality is accomplished more indirectly in the first analysis through the meticulous effort of establishing biking as a non-obligation and as a non-alternative to car driving, it is accomplished much more directly in the second analysis through the inevitable moral obligation to drive the car one already owns. Importantly, then, I do not suggest that the analyses show that automobility is maintained because they show that the participants are car owners and that they only ride a bicycle occasionally. Strictly speaking, my analyses do not concern the material ownership of cars or whether or not Tine rode a bicycle to work two or three times a week. Rather, I suggest that they demonstrate that it is possible to accomplish and utilise car ownership and biking in clear automobility-maintaining ways in rational and moral accounts of one's transportation conduct.

In terms of recent studies of governmentality, the analytical findings in Chapter Six have shown that the governing of automobility is currently unfolded at the molecular level, namely at the level of accounting for conducting one's own transportation conduct. Firstly, I would like to suggest that in relation to the observation that current forms of governing are bound up with a specific utilisation of the freedom of the governed, it is notable that in the first analysis, biking is rationalised as a free choice, and in the second analysis, the obligation to go by car is invoked to account for an already maintained practice. Thus, it might be possible to point to a general feature of the obligations invoked in the first analyses and, seemingly, in the excerpts referred to. In these accounts of transportation practices, the accomplished obligations were all bound to categories such as 'car owners', 'employees', 'dog owners' and 'healthy persons', and they were all utilised to account for individualised projects. Secondly, while the participants maintain automobility as they maintain their own freedom, this is, nevertheless, accomplished through a meticulous resistance to the moderator's questions.

Hence, the two analyses can be seen as meeting the challenge posed by Bröckling et al. (2011) and Malpas and Wickham (1995): to provide studies that pay attention to the fact that even though current governing might look like a smooth process in which the freedom of the governed is utilised for the benefit of the governing, the process is by no means smooth, but rather co-constituted by an ongoing resistance to having one's conduct conducted. In other words, while the focus of these first two analyses was to demonstrate the accomplishment of the conduct of one's own transportation conduct, the findings can be said to underline that one's own conduct is never conducted in a vacuum, it is always already conducted as resistance at the intersection between attempts to conduct the conduct of others and attempts to conduct one's own conduct. Yet, in this chapter, it was not readily observable how this more dispersed intersection is intertwined with the slightly less dispersed intersection of the municipal attempt to conduct citizens' conduct in the face of climate change and the participants' attempt to conduct their own conduct. In contrast, this is observable in the next chapter.

Concluding the chapter, consider that its findings are based in the detailed analysis of only single instances. However, as pointed out elsewhere (Lindegaard, 2012), both 'biking as a non-obligation' and 'the inevitability of car driving' appear to be utilised quite often in the larger corpus of the study. Considering biking, there are, for example, numerous instances in which biking is treated as a means of exercising and as a means of enjoying/seeing different places. In all examples, biking seems to be treated as an opportunity, not primarily for transportation, but for something else, and, in contrast, nobody ever seems to accomplish biking as something that is related to one or another category-bound obligation to get somewhere; for instance, nobody accounts for biking by saying, 'you know, I need to go to work', 'I need to go to the shop' or 'I need to visit my elderly parents'. Considering the inevitability of cars and car driving, there are numerous other instances in which the 'car owner' category is made observably relevant: when the participants talk about their first cars, when they talk about the brands of their cars, and when they invoke their car ownership in connection with the topics of shopping and going to work. Concerning the inevitability of the 'car owner' category in relation to environmental issues, the display of inevitability might be even more obvious. For instance, one participant displays the inevitability of the category when he answers the moderator's question about whether he considers environmental issues in connection with driving his car. He says that the issue concerns him and that he has bought 'one' that constantly tells him how much petrol he is using. Later, it becomes clear that 'one' refers to the specific type of car he has, making it clear that his membership of a 'car owners' category is the observably relevant starting point of his account. Another participant answers a similar question by saying that she 'of course' considers the environment, but that 'you' also have to consider what you can afford, and that it can easily cost a couple of hundred thousand kroner to *change* cars.

Note, however, that I am here concerned with a slightly different question but the ordinary question of generalizability. Considering the ordinary question of generalizability, the concern is whether a particular practice is more or less likely to be generalizable to other settings.[17] Considering, in contrast, the most prominent concern of this book, it is rather the observation that the availability of certain resources is in itself reason enough that we could and

17 As pointed out by Peräkylä (2004) and Sacks (1984: 22; 1992a: 485), the strict, ethnomethodological answer to this question would be that if a given practice occurs, then it is an undeniable part of culture and implies, as such, the potential for happening again. See also Rawls (2005: 165 and 187–8) for a discussion of the notion of generalizability from an ethnomethodological perspective.

should begin questioning what society is like if this resource is available; that is, the concern of this chapter is the observation that when the citizens negotiate the call to change their transportation practices in a certain way, it *indexes* an availability of particular resources co-constituting the regime of automobility (Rapley, 2001: 318).[18]

18 Furthermore, the book is of course concerned with the theoretically based outline of an approach that affords the possibility to study how certain rationalities of the regime of automobility are made relevant and maintained in the local discursive interaction. See Silverman (2013: 146) for a discussion of theoretical generalizations.

7. The negotiation of a transportation strategy

Whereas Chapter Six shows how the rational, moral order of one's own transportation conduct is co-constituted by two observably relevant features, namely, on the one hand, biking as a *non*-obligation and, on the other hand, an indisputable 'car owner' category, Chapter Seven points out how the parties to the focus group setting accomplish the contested intersection of the municipal strategy for 'improving' citizens' transportation practices and citizens' self-administration. Here, I focus on how the parties to the setting utilise the machinery of conversation to negotiate what the municipal strategy aims at and to what extent this makes the participants accountable for conducting their transportation conduct in a specific way. For instance, I demonstrate how the municipal strategy is accomplished with the aim of 'greening' citizens' driving practices and how this call to change one's conduct is skilfully negotiated, in that the participants ascribe themselves to an 'elderly' category, thereby accomplishing their own driving as already being as green as possible. Hence, Chapter Seven shows that in and through the pre-emption of the attempt to conduct their conduct, the participants co-accomplish the governmental strategy that thrives on the success of utilising the participants' freedom. Hence, following Bröckling et al. (2011: 18), this chapter highlights a performative relation in which governmental strategies and resistance encounter and define each other in the continuous project of governing automobility in the face of climate change. In other words, the starting point of this chapter is the understanding that the relationship between such strategies of power and resistance is inherently contradictory, resistance marking both the boundary and the constitutive moment of government.

The following analyses all demonstrate how the municipal strategy is negotiated and accomplished as a proper strategy aimed at properly conducting the proper transportation conduct of others. In the first analysis, this happens as rational accountability for the strategy is transferred to or displaced on the incumbents of *other* memberships. In the second and third analyses the municipal strategy is accomplished in and via pre-empting rational accountability for the strategy. Consequently, the analyses show how the obligating call of the campaign is transferred and pre-empted, and, by implication, they show how the governing is done in a way that utilises the participants' freedom.

The utilisation of 'someone else'

The following analysis demonstrates how the participants approach the municipal strategy as a strategy that is not really aimed at them, but at someone else.

The excerpt discussed in the following stems from a session carried out after the municipal theme month, concerning the participants' reaction to the municipal transportation theme month, focusing on how they related the strategy to their own practices. The interaction represented in Excerpt Three is from the very beginning of the session, and, by implication, it displays the moderator's proffer of the first topic, namely the participants' overall experience of the transportation theme month. In the interaction represented, the parties to the focus group setting transfer their rational accountability for the strategy to incumbents of other memberships in that they ascribe to memberships that are *not* accountable for knowing about the transportation theme month. Simultaneously, however, they resist, but thereby invoke, memberships that *are* accountable for knowing, namely memberships bound by predicates such as 'using public transportation' and 'going to work'. In this way, the municipal strategy is accomplished as aimed at conducting other members' transportation conduct, namely members using public transportation and members going to work. As such, the analysis does not directly demonstrate how the parties to the setting reflect on and/or justify their practices in the face of the strategy aimed at changing those practices. Rather, the analysis shows how the parties to the setting negotiate what and how much knowledge of the strategy they *should* have according to their specific membership incumbencies.

Excerpt Three

Participants: Moderator: mod. Jenny: jen. Svend: sve. Birgit: bir. Hans: han. Also present: Carin and Jakob

1	mod	i dag (.) vil jeg jo gerne snakke om (.) den today (.) i would as you know like to talk about (.) the
2		<transporttemanmåned> der var i byen transportation theme month that was carried out in the town
3		her i november (1.7) ((tjekker telefonen)) uanset om i nu this november (1.7) ((checks the phone)) whether you as it is
4		har lagt mærke til det eller ej have recognised it or not
5	mod	((small laughter))
6	mod	.hh øhm (.) å øh å så'n lidt ligesom .hh um (.) and um and somewhat like
7		vi plejer vil jeg gerne starte med å we usually do i would like to begin by
8		høre (.) jer hv- jer hv- jer hver især hearing (.) from e- each of you

9		.hh om i har (.) om i har haft nogen .hh whether you have (.) whether you have had any
10		oplevelse af at der har været en experiences of the presence of a
11		<transporttemamåned> .hh å hvis i har transportation theme month .hh and if you have
12		hvad den så'n- (.) om-om >hvordan i så what it has been like- (.) wh-whether how you
13		har- (.) opdaget det =om i har set en have- (.) realised it whether you have seen a
14		folder nede ved bageren eller om i- (.) flyer down at the baker or whether you- (.)
15		bare slet ikk' har set noget (.) om i have just not at all seen anything (.) whether you
16		har snakket med anne eller noget have talked to anne or anything
17		(1)
18	jen	((ser mod s)) (stod de ikk') om ved brugsen ((looks at s)) weren't they down at the supermarket
19	sve	jo å det har al'så osse været i- (1) yes and it has as well been in (1)
20		i (et af) dagbladene i hvert fald in one of the papers for sure
21	jen	var det ikk' i folketidende i guess it was in folketidende
22	sve	folketidende- eller [( ) folketidende or [( )
23	jen	[jeg [i
24		tror det var i ekstraposten osse guess it was in ekstraposten as well
25	sve	jeg mener jeg har [læst det i think i [read it
26	jen	[ja [yes
27	sve	i ekstraposten in ekstraposten
28	jen	det mener jeg osse jeg har i think did as well

29	sve	men det er ikk': (.) da jeg aldrig but it's not (.) since i never
30		bruger offentlige transportmidler gik use public transportation i
31		jeg ikk' så meget op i det didn't pay it that much attention
32	jen	°nej° no
33	mod	fordi det handlede om offentlige transportmidler because it was about public transportation
34		[det [what
35	sve	[ja [yes
36	mod	der var der was there
37	sve	ja det mener jeg det var yes i think it was
38	mod	ja (1) ka du- men- så du å du lagde ikk' så'n:, (.) yes (1) could you- but- so you and you didn't really (.)
39	sve	næ no
40	mod	=mærke til [yder- notice any [mo-
41	sve	[(for) [(because)
42		jeg skal jo ikk' på arbejde eller as you know i'm not going to work or
43		[noget= [anything
44	mod	[nej [no
45	sve	=som helst (jo) at all (of course)
46		[(heldigvis) [(luckily)
47	mod	[du havde- [you had-
48		du havde bare på fornemmelsen at det var you just had a feeling that it was

49 et eller andet med nogle offentlige
something about some public

50 [transportmidler.
[transportation

51 sve [ja al'så jeg
[yes well i

52 regnede med at det var en undersøgelse-
thought it was a survey

53 (.) for folk som virkelig bruger de
(.) for people that truly use the

54 offentlige- så'n fortolkede jeg det
public- that was how i interpreted it

55 mod ja:
yes:

56 (1.5)

57 mod a'så hv- så'n hvor man kunne s-f- (.)
well wh- like where you could a-f- (.)

58 svare på nogen spørgsmål eller
answer some questions or

59 mod [=et eller andet:
[something like it

60 sve [nej det var jeg
[no i didn't

61 ikk' klar over a'så [jeg v-
realise that well [i w-

62 mod [nej
[no

63 sve jeg øh jeg har jo-
i um i have you know

64 bir ((clears throat))

65 sve jeg har været for
i have been too

66 meget væk i det- dette efterår her så
much away this- in this fall now so

67 jeg har ikk' gået så højt op i det
i haven't paid it so much attention

68 mod nej:
no:

69		(.9)
70	sve	øh jeg jeg så lige overskriften (da um i i just saw the headings (when i
71		jeg sad å bladred) de- (.) gamle was sitting flicking) through the (.) old
72		aviser igennem da jeg kom hjem papers as i came back home
73	mod	ja: yes:
74		(2)
75	mod	<å der var et eller andet.> and there was something
76	sve	jae: °å så° (.) jeg skal jo ikk' på yes and then (.) i'm of course not
77		arbejde jeg skal ikk' i skole blom going to work i'm not going to school blom
78		eller noget or anything
79	han	nej det det no that's right
80	sve	nej det må være som skal bru:ge det no it has to be for those who need it
81		(.3)
82	mod	°ja° yes
83		(.9)
84	han	ja: yes
85		(1)
86	sve	hvis jeg kører nogen steder så har jeg if i'm going anywhere then i have
87		jo min cykel eller min bil jo you know my bike or my car you know
88	mod	°uhm° um
89		(.)

The excerpt comprises several parts. Firstly, in lines one to five, the moderator introduces the topic of the session. Secondly, lines six to sixteen display the moderator's topic proffer; that is, her request that they talk about this topic. Thirdly, negotiating a likely source of their yet to be defined knowledge, Jenni and Svend deliver a called for second pair part to the moderator's first in lines eighteen to twenty-eight. Lastly, in lines twenty-nine to thirty-one, Svend accounts for the appropriateness of his knowledge of the theme month. Since the moderator treats this as an opportunity for a 'second try' on the topic, the account ends up initiating a lengthy negotiation or, as it is, a sequence of sequences in which the moderator tries to reproffer the topic and in which Svend tries to isolate himself and thus prevent himself from being held accountable for knowing about the topic. For the sake of lucidity, I follow this structure roughly, going through the interaction one step of the sequence at a time, of course continually indicating how the different parts co-constitute the overall sequence.

Analysing the first part of the excerpt, I focus on how the moderator projects the topic. Particularly, I focus on how her introduction and topic proffer come off as projecting a sort of reserved or hesitant or even isolated second pair part. As demonstrated below, this happens as the moderator at one and the same time presents the campaign as *supposed* to be known, but *probably* not known. Introducing the topic in lines one and two, the moderator treats both the topic and the fact that she wants to talk about it as common knowledge. The moderator says, 'i dag vil jeg jo gerne snakke om den transporttemamåned'/'today i would as you know like to talk about the transportation theme month', and the combination of the commonplacedisplaying, epistemic adverb 'jo'/'as you know' and the definite reference to the theme month ('den'/'the') reveals that she understands the campaign in the way mentioned before. However, naming the topic (the transportation theme month), she more or less spells the word out, pronouncing it slowly and with almost equal stress on each syllable. This noticeably careful pronunciation shows that the moderator does not really expect the topic to be part of the participants' common background knowledge. Furthermore, continuing after the pause in line three (notice that the long pause in line three is to be read as the moderator's, since she displays her silence as a matter of checking her phone, not as a next speaker invitation) she says that she wants to talk about the theme month and *whether or not* the participants have recognised that it is taking place. Afterwards, in line five, the moderator laughs, and her laughter comes off as a sort of 'ambivalence device' or what Schegloff refers to as a 'negative polarity item' (Schegloff, 2007a: 173). More specifically, whereas the introduction has projected that the participants might discourage the topic, her laughter premitigates such a future dispreferred interactional situation.

The moderator's topic-proffering lines (six-sixteen) emphasise that she understands her proffered topic as somewhat and somehow 'troublesome'. The proffer comprises a relatively high amount of different features displaying uncertainty (such as self-initiated self-repair) and a number of negative polarity items displaying an orientation to a negative answer, enhancing the already invoked display of ambivalence. I go through these one at a time, beginning with the uncertainty features and proceeding with the polarity items.

Considering the uncertainty features, in line six the moderator initiates her proffer with an audible breath, an 'um', a very brief pause and yet another 'um', and in line eight she twice restarts the individuating expression 'jer hver især'/'each of you' Additionally, in line nine she restarts 'om i har'/'whether you have', and in line twelve she carries out self-initiated self-repair two times, changing 'hvad den sådan'/'what it has been like' firstly to 'om'/'whether' and secondly to 'hvordan'/'how'. Considering the negative polarity items, the just mentioned double repair in line twelve is interesting. By saying 'hvis i har'/'if you have', the moderator initiates an ordinary counterfactual construction in the second half of line eleven, and line twelve displays the fulfilling consequent of this. Due to the 'counterfactuality', it could be argued that the hypothetical construction invokes an orientation to a negative answer. Furthermore, it could be argued that the double repair in the consequent increases this orientation. In particular, line twelve is changed from a question of *what* to a question of *whether* and *how*. Hence, the first seems to project a rather direct question about what the theme month has been like; the second seems to project that there is no answer to this question; and the third seems to project something in between the two. However, while the final option may at first seem to be equally oriented to a positive and a negative answer, the subsequent exemplifying 'list' should be noted. More specifically, the moderator concludes her proffer with a list suggesting *how* the participants might have come across the campaign, and this list includes the option that the participants have noticed nothing at all (lines fourteen to fifteen).

It is time to recall my initial claim that the excerpt demonstrates how knowledge of the transportation campaign is accomplished in a way that isolates and prevents members from being held responsible for having recognised and considered it. It is of course necessary to account continually for this claim, but at this point the moderator's (first) proffer can be said to project such an accomplishment. In particular, her proffer has demonstrated how difficult it is to even ask somebody about their *reflections* on a *potentially* automobility-challenging campaign.

In the following, it is pointed out how Jenni and Svend cooperatively produce a bit of knowledge that for all practical purposes co-constitutes a layered second pair part to the moderator's first. Whereas there are quite a few brief pauses in the moderator's introduction and proffer, none of them come off as obvious next speaker invitations. Although the moderator marks the pauses in lines fourteen and fifteen as possible transition relevant places (she looks around at the participants when she talks to them), she does not treat her listing activity as complete. Thus, the longer pause in line seventeen comes off as the first invitation, and, by implication, Jenni's self-selected response in line eighteen is delivered at an unremarked point, projecting nothing but a preferred structure.[1] However, Jenni does not respond directly to the moderator. Rather, she turns to Svend and starts a sort of pre-second insert expansion.

According to Schegloff (2007a: 106), a pre-second is a type of action or sequence of actions that is, as the name indicates, introductory to the second pair part. The action is relevant in situations where it is necessary to establish some resource for delivering the second pair part made relevant by the first. As demonstrated below, Jenni's response in line eighteen initiates such an attempt to provide better resources before delivering the second proper. Whereas most pre-seconds are directed to the producer of the first pair part, Jenni's action is directed to a fellow participant, Svend. Asking Svend whether 'they' were in fact 'down at the supermarket', Jenni produces a first pair part in the ongoing question-answer sequence constituting the pre-second used for accomplishing the knowledge needed for the second proper. Accordingly, initiating his turn in line nineteen with a 'yes', Svend fulfils both the pre-second and the called for second pair part to the moderator's first pair part. Hence, by accomplishing the knowledge Jenni and he have about the campaign (because they know that 'they' were down at the supermarket), Svend simultaneously fulfils the second proper. In this way, Jenni and Svend cooperatively display a stance which preferably encourages the proffered topic.

Strictly speaking, it cannot be 'verified' that Jenni and Svend are talking about how they noticed the *theme month*, since they do not mention it explicitly. However, drawing on structure and membership categorisation analysis, it is possible to reason about it. Firstly, it should be noticed that by mentioning the theme month in her introduction, the moderator makes the topical device 'transportation theme month' relevant. Furthermore, recalling that Svend's 'yes' in line nineteen not only fulfils Jenni's pre-second, but also the relevant second to the moderator's proffer, lines eighteen and nineteen come off as relevantly

1 In a focus group setting where the moderator does not select a next speaker, it is only expected that there will be a longer pause before someone self-selects than what is common in most so-called naturally occurring talk-in-interaction.

oriented towards the proffer. Accordingly, drawing on Sacks's consistency rule, Jenni could be heard as invoking one or more membership categories from this particular topical device. For instance, her 'they' in line eighteen could (and should, according to the economy rule) be heard as belonging to some or other membership connected to this device. Hence, at a more refined level of analysis, it can be said that Jenni and Svend accomplish knowledge about members belonging to the topical device.

As mentioned, the interaction represented in lines eighteen to twenty-eight displays the embedded sequence which makes up the second pair part and which is made relevant by the moderator's first. Thus, it was not really warranted to treat Svend's 'yes' in line nineteen in isolation. However, this semantically motivated treatment of the accomplishment of Jenni's pre-second served the analytical purpose of highlighting the structure of Jenni and Svend's cooperation. Regardless, recall that Svend does not in any sequential way mark his 'yes' in nineteen as a finite turn but rather treats the 'yes' as the beginning of a round, locating the social visibility of the campaign. In particular, he adds that the campaign might also have been mentioned in one of the papers, and it should be noted that he and Jenni cooperate on establishing in which paper they have read about 'it', once again clearly exemplifying how knowledge is accomplished in situ for all practical purposes. Jenni suggests that [she has read about the theme month] in 'folketidende' in line twenty-one. However, by breaking off, pronouncing 'folketidende' and continuing with an 'eller'/'or', Svend shows that he is not sure (line twenty-two). Jenni willingly participates in the anticipated search, immediately overlapping Svend's 'eller'/'or', saying that she 'tror'/'guesses' it could be [that she saw something] in 'extraposten'. In lines twenty-five and twenty-seven, Svend agrees; he also 'mener'/'thinks' he read about it in 'extraposten', and in line twenty-eight, Jenni again says that she 'mener'/'thinks' she did so as well.

In this way, Jenni and Svend establish that they have read something related to the theme month in a specific paper. Furthermore, they treat the vague modality of this establishment as appropriate for all practical purposes, namely for delivering *some* knowledge about the transportation theme month. Yet, while the participants do not treat their way of accomplishing knowledge as in any way problematic, they do treat it as a matter of important negotiation how much knowledge about the transportation theme month they can be held accountable for.

Initiating his turn with the turn construction unit 'men det er ikke'/'but it's not', Svend marks a shift in line twenty-nine. Seeing as the conjunction can be heard as marking a distance to the preceding interaction (which was treated as preferred), the unit can be heard as projecting a dispreferred action and, as it is, the subsequent interaction does have a typical dispreference structure. In more detail, lines twenty-nine to thirty-two present an account delivered through a conditional structure. Through the antecedent 'da jeg aldrig bruger offentlige transportmidler'/'since i never use public transportation' and the consequent 'gik jeg ikke så meget op i det'/'i didn't pay it that much attention', the lines display an accounting mitigation of the just accomplished knowledge about the transportation theme month. Thereby, Svend's turn initiates an important negotiation of the otherwise established proffered topic. More specifically, the turn initiates a process in which Svend tries to isolate himself from the moderator's repeated attempts to elaborate on her proffered topic, namely the transportation theme month.

Before starting the categorial analysis, I provide a few comments on the structure that co-constitutes it. By displaying a discouraging extension of the second pair part, Svend's utterances in lines twenty-nine to thirty-two mitigate or even undermine the so far established alignment of the second. In general, dispreferred second pair parts are expansion relevant, because of the normatively called for attempts to better alignment. However, even though the present sequence is also continued, it is continued in a way that is typical, especially for topic-proffering sequences, though not for sequences comprising dispreferred seconds in general. Again, when Schegloff (2007a) refers to the 'organizational contingencies' of topic talk (158), he implies that post-second activity in topic-proffering sequences is not to be considered ordinary post-expansion, but rather some sort of 'sequence(s) of sequences'. This is most often the case after a preferred second projecting topic elaboration, but as the topic-proffering speaker often provides 'second tries', it can also occur after rejected proffers. The latter is exactly what happens in the present interaction. In line thirty-three, the moderator delivers the first in a range of 'second tries', all of which, as demonstrated below, in different ways negotiate the use of Svend's in-between pieces of category work.

Accordingly, reconsidering Svend's account in lines twenty-nine to thirty-two, its co-constituting category work should now be considered. Firstly, in the antecedent of the conditional clause (that is, in 'since i never use public transportation'), the predicate 'using public transportation' is made observably relevant. Again drawing on Sacks's consistency rule, it is reasonable to hear the predicate as bound to a membership category belonging to the already accomplished topical device 'the transportation theme month'. The membership is invoked as Svend *resists* its bound predicate and, by implication, the membership as such. Secondly, in the consequent (that is, in 'i didn't pay it that much attention'), some sort of category-bound right or obligation is invoked. This happens when Svend infers from his *non*-membership of categories bound by the predicate 'using public transportation' that he did not pay much attention. In an ordinary logical hearing, the antecedent is heard as justifying or grounding the consequent, and, by implication, Svend's non-membership is treated as justifying his lack of attention.

Hence, Svend resists membership of a category bound by the predicate 'using public transportation' and by the obligation 'to pay attention to the theme month'. However, when he resists the categorisation, he also invokes it, thereby projecting a group of persons who use public transportation and are *specifically* accountable for knowing about the transportation theme month. Further, before Svend's category work, no one has mentioned public transportation. Accordingly, Svend's work comes off as a specifically reached for contribution to the ongoing accomplishment of the theme month. That is, he specifically reaches for an understanding/notion of the month which mitigates his obligation to know about and reflect on it.

Jenni's acknowledging brief response in line thirty-two displays an ordinary minor post-expansion, doing the job of a sequence-closing third (Schegloff, 2007a: 118). However, responding in line thirty-three, the moderator does not recognise this possible closing. In particular, as she does not treat the preceding sequence as complete, she does not accept her topic proffer sequence as rejected, failed and closed. On the contrary, she in fact delivers a 'second attempt' to proffer the topic. She does so in the form of a request, asking Svend whether his inference from not using public transportation to not paying attention to the theme month campaign was caused by an understanding of the campaign as being focused on public transportation. Thereby, she negotiates his account work, utilising it for her own purposes. Rather than orienting to Svend's consequent, accounting for not knowing, the moderator suggests that his antecedent utterance might make a specific presupposition observably relevant. She suggests that his account may after all imply that he knows something about the theme month, namely that it focused on public transportation. Hence, the moderator neglects Svend's account work; rather than fulfilling an account of not knowing, she accomplishes a piece of knowledge.

However, Svend responds to the moderator's question via two 'yeses', one in line thirty-five and one initiating his turn in line thirty-seven. At first, these responses seem to display preference; the first is straightforward and occurs in a transition relevant place (in inadvertent overlap), and even though the modality of the second 'yes' is slightly degraded in the rest of line thirty-seven (he says, 'ja det mener jeg det var'/'yes i think it was'), there is nothing to indicate dispreference. Yet, recalling that with regard to topic-proffering sequences, preference is primarily a matter of a stance that encourages elaboration of the proffered topic, Svend's 'yeses' could be heard as predominantly *closure* relevant. Such an understanding is substantiated in the display of disalignment in the following interaction.

The following interaction comprises three 'blocks', all marked by actions by the moderator. Firstly, in lines thirty-eight and forty, the moderator delivers a 'second attempt' to which Svend responds with dispreference in the following lines, until line fifty-four. Secondly, in lines fifty-seven to fifty-nine, she delivers yet a 'second attempt', and once again Svend responds, displaying dispreference. Lastly, in line seventy-five, the moderator finally acknowledges that she will not get more from Svend on the topic concerned, and the remaining part of the excerpt represents how not only Svend and the moderator, but also Hans cooperatively close Svend's second pair part contribution.

Beginning with the first of these blocks, the moderator's turn in lines thirty-eight and forty reveals several turn construction units and perhaps more than one turn. Whereas her 'yes' and the pause in line thirty-eight is not readily heard as anything more than just an acknowledgement of Svend's preceding turn, she and Svend cooperatively display that the interaction proceeding the pause is a 'second attempt' at topic proffering. After the pause in line thirty-eight, the moderator self-initiates and self-repairs her potentially more direct request ('kan du'/'could you'), reframing it on a negative polarity item ('ikke sådan'/'didn't really'). The negative polarity displays an orientation to a negative answer, and in an ordinary sequential environment, the moderator could, as it is, be heard to project closure. However, again recalling the particularities of topic-proffering sequences, it is not too surprising that Svend does not fulfil the closure but rather treats the moderator's work as yet another topic proffer, delivering an account.

In an ordinary sequential environment where the moderator was heard to project closure, a closure relevant response would of course have been preferable. Considering this, because Svend gives a dispreferred response, it appears to treat the moderator's work as a topic proffer, not as a closure initiator. Hence, Svend's response apparently demonstrates a subtler orientation to preference and dispreference than these rather dichotomous categories might themselves suggest. By delivering a dispreferred response, he at one and the same time acknowledges his local membership of 'parties to a focus group setting' and displays that his stance towards the topic remains discouraging all the same.

At one and the same time aligning with the overall 'doing focus group' business and disaligning with the more specific project proffered by the moderator, this subtle account is delivered in lines forty-one to forty-six in a perfectly transition relevant place, pre-empting the moderator's assumed last phrase. The perfect timing shows that Svend acknowledges his membership of 'parties to a focus group setting', drawing on it to negotiate the premises of the moderator's project. Particularly, it reveals how he draws on the opportunities provided by this membership to deliver yet another piece of justifying and negotiating category work.

In lines forty-one to forty-six, he says, 'for jeg skal jo ikke på arbejde eller noget som helst jo heldigvis'/'because as you know i'm not going to work or anything at all of course luckily', thereby resisting membership of a category bound by the activity 'going to work or anything'. Again drawing on Sacks's consistency rule, it is reasonable to hear this category work as being oriented towards the topical device of the transportation theme month. Resisting categorisation, Svend makes the category an observably relevant resource in his project of isolating himself from the theme month. Yet, he does so precisely when he ascribes the categorisation and, by implication, the accountability to others: to people who are, on the other hand, bound by the predicate 'going to work'. Hence, whereas Svend's first isolation work was done via his resistance to a membership of persons who are 'using public transportation', this part is done via his resistance to a membership of persons who are 'going to work'.

The moderator responds to this category work in lines forty-seven to fifty, treating it as mitigating what Svend can be expected to know about the subtopic *public* transportation. In particular, suggesting that Svend just had a *feeling* ('fornemmelse', line forty-eight) that it had to do with public transportation, she approaches the category work as mitigating his accountability, and nothing in her turn points towards further attempts to reproffer the topic; for instance, it is ended with an audible downwards intonation.

Svend, however, does not treat it as an opportunity to withdraw himself from the interaction. On the contrary, he displays an interest in establishing the ongoing understanding of the topic as something that someone like him cannot be expected to know about. Furthermore, since the first part of his turn in line fifty-one overlaps with and pre-empts the last bit of the moderator's turn in the same line, he reveals that he finds such an establishment quite important. The interaction thereby initiated displays yet another piece of category work, but not like the preceding; it is done via direct ascription, not via resistance. Broadly glossing, in lines fifty-one to fifty-four, Svend suggests that the theme month was unfolded through a survey aimed at people who truly rely on public transportation. He thereby invokes membership of people bound by the predicate 'to truly ('virkelig') use public [transportation]' (lines fifty-three and fifty-four). Ascribing accountability to another group, he once again resists personal accountability. Furthermore, the stressed and superlative-like pre-modifier 'virkelig' comes off as reached for or chosen in particular. Strengthening the bound relation between category and predicate, the pre-modifier shows that the predicate is no mere random predicate but rather is an obligating, category-bound activity. Hence, the stress on 'virkelig' fine-tunes the isolation work. The stress limits the scope of accountability to members obligated to use public transportation as a *category*-bound activity, thereby isolating Svend from accountability, even if he *happens* to use public transportation once in a while.

Similarly, the moderator does not approach Svend's category work as being primarily an isolation resource. In contrast, she again negotiates the practical purpose hereof, utilising it as affording yet a second try. Hence, a number of points should be observed. Firstly, responding with a 'yes' and a clearly questioning intonation, the moderator displays orientation towards Svend's work, anticipating elaboration, *not* isolation from being held accountable for such elaboration. Secondly, even though her 'yes' is met merely with a rather long pause, potentially projecting a dispreferred response, she continues in line fifty-seven, pursuing the elaboration work she requested. While the moderator does display orientation to the notion that she might be pursuing a dispreferred project, beginning her reproffer with a hedging 'altså'/'well' and self-initiating and self-repairing twice in the first line, she nevertheless pursues the topic talk. In more detail, she pursues the request for elaboration displayed in her questioning 'yes', asking whether the survey Svend mentioned was concerned with answering questions or the like. Thus, it should be recognised that whereas Svend draws on the mention of 'survey' to resist being held accountable for considering the theme month, the moderator draws on the mention of it to pursue Svend's experience of the theme month.

Svend's response fulfils the dispreference projected by the moderator's slight hesitation. Particularly, it should be noted how he reveals that he disagrees with the premise of the moderator's question. Firstly, his turn in line sixty is initiated with the negative and type-conforming token 'nej'. Secondly, even though the 'no' overlaps in a perfectly transition relevant place (once again pre-empting the last part of the moderator's turn), it can be heard as delayed because of the rather long pause not taken in line fifty-six. Also, Svend's answer proper, his disconfirmation, is substantiated through yet another justifying account. In the middle of line sixty-one, Svend prefaces the account with a hedging 'well', in lines sixty-one and sixty-three he restarts and self-repairs and, finally, the account as such is delivered in lines sixty-five to sixty-seven.

Again, the account is unfolded via a bit of category work. However, this category work is different from the former in that it does not draw on the direct or indirect resistance to activity predicates (like the former 'using public transportation', 'going to work' and 'obligated to use public transportation'). Contrastingly, predicating himself as 'having been too much away this fall' (lines sixty-five to sixty-six), Svend draws on a location predicate as his justification for not having paid that much attention (a consequence he explicates in lines sixty-six to sixty-seven). Accordingly, since Svend draws on a resource which clearly is applicable only to him, this category work is notably more individualistic than the previous. Hence, while it might be possible for the other participants to cast themselves within the already suggested knowledge-resisting categories, the presently invoked location resource is presented as available only for Svend. That is, all the parties to the focus group setting know that Svend has been travelling in Eastern Europe in the fall and that this, among other things, has influenced the planning of the present focus group session.

In line sixty-eight, the moderator responds to the account with an acknowledging 'no', and after a pause, Svend proceeds in lines seventy to seventy-two, elaborating on the account. He points out that he, being at home again, leafed through the *old* papers ('gamle aviser', lines seventy-one to seventy-two). Following this, it could be suggested that he does not merely utilise the location account to *withdraw* himself from accountability, but also to display the fact that he knows anything at all about the papers' mention of the theme month as being of *specific* worth.

Lines seventy-three and seventy-four come off in exactly the same way as fifty-five and fifty-six. Hence, again the moderator's 'yes' is questioning, and the pause represents a next speaker invitation. Again, Svend does not take up the invitation projected by the pause. However, when the moderator proceeds in line seventy-five, it is in a remarkably different way than when she herself proceeded, pursuing the elaboration in line fifty-seven. She displays no attempts to circumvent the purpose of Svend's account. On the other hand, using a settling intonation, she suggests the possible understanding that perhaps Svend just happened to see something or other ('der var et eller andet'). In line seventy-six, Svend confirms this understanding, and he immediately uses this common ground as a starting point for settling the previous negotiation for his practical purposes. This happens in a number of steps.

Firstly, after confirming the moderator's candidate understanding with a 'yes', Svend elaborates and repeats one of his already given accounts. More specifically, in lines seventy-six to seventy-eight, he repeats the 'I'm not going to work' account he provided in lines forty-one to forty-six. This time, however, he upgrades the account, adding, 'jeg skal ikke i skole blom'/'i'm not going to school blom' and looking at Hans Blom. Secondly, after Hans has delivered an agreeing response in line seventy-nine, Svend repeats another of his accounts, namely his 'people who are obligated to use public transportation' account. Again, in lines eighty-two and eighty-four, both Hans and the moderator provide brief acknowledging responses. Lastly, Svend could also be heard as repeating his first account, that is, the account in which he resisted the more general predicate 'using public transportation'. In lines eighty-six to eighty-seven, Svend characterises his knowledge resistance as being a matter of his membership of people who own a bicycle and a car, thereby emphasising that the campaign was not for him, but for 'others'. This last classification is unfolded when he presents it as a necessity that if he goes anywhere it should be acknowledged that he has a bicycle and a car, thereby invoking if not an owner's obligation to ride a bicycle or to go by car, then at least a right to do so.

In this way, Svend treats the moderator's acceptance of his location account as an extension of his entire account work. Hence, in the end, all of Svend's invoked categories are treated as doable resources for isolating and preventing him from being held accountable for the transportation theme month. A specific understanding of the theme month is thereby made relevant. Particularly, in resisting knowledge about the theme month, the theme month is invoked as a specific matter, namely as a strategy for people using public transportation and/or going to work and/or school, but *not* for people who own a bicycle or a car.

While the purpose of this analysis has been to show how the participants transfer accountability to other people, the purpose has not been to display 'transference' as explicitly the participants' agenda. Rather, this practice of transference has been an element in subtle negotiations. Primarily, the participants, notably Svend, attempt to isolate and prevent themselves from being held accountable for knowing about the municipal theme month. However, as part and parcel of this task, certain other membership incumbents have been invoked for the practical purpose of transferring accountability. Hence, two types of memberships have been invoked: the non-accountable and the accountable. Due to the individual features of these memberships, the transportation theme month has been accomplished in a specific way, namely as a theme month on public transportation and on commuting to work and school. Accordingly, while the transference of accountability does co-accomplish the municipal strategy as a rational governmental attempt to conduct citizens' transportation conduct in the face of climate change, it simultaneously accomplishes the (limited?) orientation of this rational attempt to govern automobility.

The utilisation of category membership

This analysis and the next both feature analyses that show how the participants pre-empt the municipal attempt to conduct the conduct of the citizens. However, the two analyses differ in that the pre-emption is done in different ways. In this section, the analysis shows how the participants approach the municipal attempt to conduct citizens' transportation conduct as not really necessary, since the participants already comply with the suggested conduct due to their specific memberships. On the other hand, the next analysis shows how the participants approach the municipal attempt to conduct citizens' transportation conduct as inconsequential, seeing as the participants already hold the relevant knowledge about the suggested conduct. In the preceding analysis of the transference of accountability, transference accomplishes the municipal strategy as a rational governmental attempt to conduct citizens' transportation conduct in the face of climate change. Similarly, the pre-emption of accountability fulfils the municipal strategy as a rational *governmental* strategy. However, neither in this analysis nor in the next does this happen in a straightforward way and without negotiation.

The excerpt discussed in this analysis stems from the same session as the excerpt discussed in the preceding analysis. Again, the session took place after the municipal transportation theme month, and it concerned the participants' reactions to the theme month, focusing on how they related the strategy to their own practices. The interaction represented in the excerpt is taken approximately six minutes into the session, and, as it is, there are only three minutes of interaction between the interaction represented here and the one presented in the above analysis. Consequently, the parties to the setting are still in the business of accomplishing a round prompted by the moderator's proffer of the first topic. Only, in between this excerpt and the preceding, also the second question concerning possible experiences of the transportation theme day (which was part of the transportation theme month) has been made briefly relevant as part of the general topic talk on (lack of) observations of the strategy. In the interaction represented in Excerpt Four, the parties to the focus group setting accomplish the in situ 'prefulfilment' of the municipal strategy concerning greener transportation practices; in particular, greener driving practices. Throughout the interaction represented in the excerpt, *green driving* is accomplished as advisable. However, the point of the analysis is to demonstrate that this is done in a remarkable way. Hence, the analysis points out how the participants ascribe to an 'elderly' category, thereby accomplishing their own driving as *already* being as green as possible.

Excerpt Four

Participants: Moderator: mod. Jakob: jak. Birgit: bir. Hans: han. Svend: sve. Carin: car. Also present: Jenny

```
1    bir   det der ikk' nogen der skal mer her
           there's no one here that has to do that anymore

2    jak   nej
           no

3    bir   [((looks at mod)) jeg har=
           [((looks at mod)) so neither i=
```

4	han	[((looks down at his cake plate
5	bir	[så heller ikk' set noget [have seen anything
6	han	[nej men det der- [no but that-
7	bir	om det i avisen (jeg ved slet ikk' about it in the paper (i don't know nothing
8		noget om det) about it)
9	mod	nej (.5) nej no (.5) no
10	bir	=jeg har ikk' lagt mærke °til noget° =i haven't noticed anything
11	mod	nej no
12		(.9)
13	han	det der kør grønt kursus (.9) det øh the green driving course (.9) that um did
14		bed jeg osse mærke i fordi det ku- i also take special notice of because i'd
15		(1.5) m- >det ku jeg godt tænke mig< (1.5) um- i'd like to
16		(1) å lære (1) learn that
17		(1)
18	mod	ja: yes
19	han	for jeg har aldrig nogensinde været øh because i have ever never been a um
20		nogen (.) hundrede procent god chauffør a (.) hundred percent good chauffeur
21		[(2.2)
22	mod	[((shakes head))
23	han	selvom jeg jo gerne vil bilde mig det although i'd of course like to tell myself
24		ind å [i- otherwise and [you
25	(mod)	[uhm [um

26	han	=i er (.) velkommen til at sige årh you're (.) welcome to say oh but
27		det er ikk' så galt that's not so bad
28	mod	[((laughs))
29	sve	[((laughs))
30	mod	[.hhh ha-ha
31	car	[årh men du kører [oh but you drive pretty
32		ganske udmærket ( ) decently
33	han	((looks at mod))
34	mod	°ha-ha°
35		(1.1)
36	car	å [ikk' f- and [not too f-
37	han	[((looks at mod)) til en afveksling [((looks at mod)) for a change
38	car	å ikk' for stærkt and not too fast
39	sve	((looks at han)) =i din alder blom der er du grøn ((looks at han)) at your age blom you are a green
40		chauffør driver
41	car	[=ja [=yes
42	sve	[du] ræ:ser jo ikk' af sted [you] know you're not rushing ahead
43	car	°ja° yes
44		(1.5)
45	han	hvad gør jeg: what do i do
46	sve	du ræser jo ikk' derudad [så er du- you're not rushing ahead [then you're-
47	car	[nej [no

48 sve [så] kører du stille å roligt
[then] you're driving quietly and easily

49 car [ja
[yes

50 sve [så-
[so

51 car [ja
[yes

52 (1)

53 han ja
yes

54 sve =så kører du så grønt som du kan vel
=then you're driving as green as you possibly can i think

55 (1.1)

56 han jo det er rigtigt (.) men al'så (.)
yes that's right (.) but well (.)

57 øh: (.) det her med (.) å (.) køre
um (.) this thing about (.) to (.) drive

58 så'n at det er <økonomisk> å så'n
in a way that is economical and so on

59 det er- (.) det vil jeg faktisk meget
that's- (.) i'd actually really like

60 [gerne
[that

61 car [(men)] det har du da altid gjort
[but surely you have always done so

62 sve °ja° ((looks at car))
yes ((looks at car))

63 (1.5)

64 sve men så ((looks towards han)) er det du skal ha ((rækker
but then ((looks towards han)) it is what you must have ((reaches

65 hen å rører han på armen)) (.) en ny bil
over and pads han on the arm)) (.) a new car

66 (.) å en computer der fortæller dig
(.) and a computer that tells you

67 hele tiden hva den-
all the time what it-

68 (1.3)

69 han [ja
[yes

70 sve [( )
[( )

71 car [( )]
[( )]

72 købe så'n en- en gps der
buy such a- a gps there

73 står der osse på [hvor stærkt du
it also says [how fast you

74 sve [gør der det

75 [does it really

76 car =ja det gør der
=yes it does

77 sve så (1.3) °skal du bare kigge på den
then (1.3) you just have to look at that

78 så° >jo det siger- det siger sig selv
then yes it's of course evident- it's evident

79 (.) når først den går et par å tyve
(.) when it first reaches some tventy

80 kilometer på literen så ka den- (.)
kilometres a litre then it can't- (.)

81 så ka det [ikk' blive
then it [can't be

82 car [jamen det
[well that

83 sve mere grønt
any greener

84 car det ka vores bil ikk' gøre
our car can't do that

85 sve (nåeh) ka [den ikk'
(oh) can't [it

86 car [nej
[no

87 car nej det ka den ikk'
no it can't

88 (2)

89 han ja så skal det være sommer
well then it has to be summer

90 car jamen det ka den ikk' (.5) den ka højst
well but it can't (.5) it can at most

91 køre atten ((ryster hovedet let))
go eigtheen ((shakes head lightly))

92 sve atten ja ja-ja (.) °(men det .hh=)°
eighteen yeah yeah-yeah (.) (but that .hh)

93 han =men det da os pænt
but that's also quite good

94 sve (det os) det virklig flot (.) men det.
(that's also) that's really good (.) but it

95 det- det tænker man over når man har-
it- it makes you think when you have-

96 a'så står der å skifter he- ah nu går
you know is sitting there alternating he- oh now it only

97 den kun ni hvis man kører op ad bakke
goes nine if you're driving up a hill

98 så gir den lidt ekstra ikk'
then it gives a bit extra right

99 han ja
yes

100 sve så det ligesomom (.) (filer lidt tilbage)
then it's like (.) ( )

101 ser hva jeg har- de to sekunder det
realise what i have- the two seconds it

102 tar mere ikk' os.
takes more right

103 (1.1)

104 mod har- har du, (.)
have- have you (.)

105 [du

106 [you

107 sve [ja
[yes

108 mod bruger du kører efter dine instrumenter.
do you drive accordingly to your dashboard

109 sve det gør jeg.
that i do

In between Svend's work discussed above and the present excerpt, Jenni and Jakob followed Svend's lead and utilised the readily at hand membership of 'elderly people' as a resource for justifying why none of them know much about the municipal strategy either. In between Jenni and Jakob, Carin also responded, but while her response was similarly dispreferred in discouraging the topic (she also disclaimed the resources for talking about it), it did not comprise any account suggesting that she *should* not know.

This excerpt begins when Birgit self-selects and delivers a response to Jakob's just delivered self-predication as someone who does not go to work anymore. Hence, it is clear that the interaction represented displays an ongoing accomplishment of a round in and through the structural pattern Schegloff refers to as multiple second pair parts. Consequently, the different second pair parts are second pair parts in that they display actions which are relevantly oriented to the first pair part; that is, to the moderator's topic proffer. Furthermore, as noted, the second pair parts have all been dispreferredly (though not less relevantly) oriented to the proffer. As discussed below, Birgit's response also comes off as dispreferred, but my main motivation for discussing it is that it represents the intertwined transition to Hans's contribution to the overall cooperative topic-proffering and topic talking sequence. Hans's contribution is particularly interesting, in that it displays an attempt to *encourage* the proffered topic which is, however, treated as a relevant problem that calls for careful negotiation. Hence, Hans displays a challenge to an otherwise somewhat stable accomplishment of the theme month as being something the participants cannot be held accountable for knowing about, and the interaction that grows from this challenge refines, explicates and adds to the structures and categories used to resist and manage the topic.

The excerpt displays an intertwined structure with lots of layered negotiation work, necessarily requiring long-winded explanations as the analysis progresses in the aim of unpacking the co-constituting category work. At this point, however, I limit the structural overview to an absolute minimum and merely point out what can be said to represent the four parts of the analysis. Firstly, lines one to eleven comprise Birgit's second pair part work. Secondly, lines thirteen to twenty-seven comprise Hans's second pair part work, and this part demonstrates how Hans's orientation towards a 'green driving' course projects a challenge of the otherwise established collective isolation. Thirdly, the interaction represented in lines thirty-one to sixty displays both a challenge of Hans's second pair part and Hans's attempt to defend the former. Fourthly, the remaining part of the excerpt displays yet a challenge, this time a somehow radicalised challenge of Hans's defence, concerning the categorical implications. The last part of the excerpt shows how Svend utilises Hans's defence as a starting point for reframing the topic that significantly provides Svend with the opportunity of orienting towards a new isolation category, namely a category of persons who already hold all relevant knowledge. Of course, I return to, qualify and clarify these parts as I proceed.

As stated, I begin with Birgit's work in the first lines of the excerpt. Birgit's work comprises the fulfilment of the ongoing collectivisation of the participants' isolation work. Since the work marks the transition to Hans's work, it anticipates the interaction which is the main focus of this analysis. Before this excerpt, Jakob said that he does not work anymore. He added that those who do should probably be the ones to 'decide on these matters'. As he did not specify this any further, it is not possible to explicate which 'matters' present-day employees should decide on, but he is probably referring to matters invoked by the transportation theme month. On the other hand, what can be explicated is that Jakob thereby gives his version of the 'I'm retired' account. Thereby, he not only classifies himself as retired (by further ascribing any decision responsibility to people who work), he also negatively exempts his own category incumbency from such binding obligations. Hence, once again the 'I'm retired' account is utilised as a predicate, invoking the 'stage of life' device and the 'elderly' category as interactional resources.

Jakob's 'membershipping' was followed by a fairly long pause. Then, in line one of the present excerpt, Birgit self-selects. Saying, 'det der ikke nogen der skal mere her'/'there's no one here that has to do that anymore', Birgit not only accepts Jakob's category work, she extends the scope of the 'elderly' membership to all the participants.[2] So far, the participants have ascribed to the category individually, but by pointing to the fact that none of the participants belong to 'stage of life' categories bound by the predicate 'has to go to work', Birgit indirectly upgrades the already accepted category 'elderly'. So far, the category has only been individually applicable, but by collectivising the inapplicability of the predicate ('to go to work') to the group as such, Birgit makes a collectivised non-worker category an observably relevant membership incumbency for all the parties to the focus group setting.

Considering the sequentiality, the moderator is *not* included in this isolated collectivity. Through the multiple second pair parts design, the members to the setting reflexively accomplish a context in which the second pair part providers constitute a collectivity of proper speakers on the topic as being somehow

2 As Hester and Eglin (1997) mention, it has since Sacks' founding work on membership categorisation analysis been suggested that collectivities (as well as non-personal objects) can be seen in the same way as personal memberships: an insight, they point out, that was, however, already anticipated in Sacks' work (3).

opposed to the first pair part provider, the moderator. Accordingly, when Birgit says 'det der ikke nogen der skal mere her'/'there's no one here that has to do that anymore', *no one here* points out this already sequentially established collectivity. Hence, note the adverb's extensive and inherent indexicality: 'her'/'here' refers to more than just place and setting, it refers to residents, consequently constructing and excluding the moderator as a non-resident, a visitor.

Bearing in mind that the 'elderly' incumbency has so far been used as a resource for mitigating dispreferred and topic-discouraging second pair parts, Birgit's turn comes off as projecting and mitigating a dispreferred response. She fulfils the projection in the base second pair part in lines three, five, seven and eight, revealing that she cannot claim the resources to sustain the topic talk. The base is simply, but effectively tied to and co-constituting the category work she has already performed using the small word 'så' in line five. The explaining and thus substantiating adverb underlines that due to her 'elderly' incumbency, Birgit cannot be held accountable for having noticed the transportation theme month. More specifically, the adverb reinforces the co-operated logic that if you are retired, then you cannot be expected to have noticed the theme month. The moderator accepts Birgit's response (and its incorporated, collectivised account), as she aligns with the polarity of it in a couple of minimally constructed turns in lines nine and eleven.

Accordingly, Birgit's second pair part work could engender closure for the overall topic-proffering sequence. By extending the account category 'elderly' to the whole setting, Birgit could be heard as being projecting sequence closure.[3] Particularly, since she has in a way already accounted for the fact that any future second pair part would be similarly dispreferred, she can be heard to pre-empt any future seconds. As indicated, also the moderator's minimally constructed acceptances show that the topic could be closed; that it is not really something to talk about in this specific setting, because of the participants' specific incumbencies. However, the possible sequential settlement and, by implication, the well-established isolation are challenged by Hans's turn in line thirteen. In the following, I demonstrate how this grows from the suggested relevancy of the course in 'green driving' previously mentioned by the moderator.

Since Hans's challenge of the collective isolation comes off through his orientation to the 'green driving' course previously mentioned by the moderator, I begin the analysis by demonstrating how it comes off as a reference to the above interaction. Hans self-selects in line thirteen and delivers the turn construction

---

3 Recall Schegloff's (2007a) observation of the systematic reversal of the ordinary differential expansion relevance of preferred and dispreferred second pair parts (169).

unit 'det der kør grønt kursus'/'the green driving course'. By treating the reference 'det der'/'the' as anaphoric rather than deictic (there is nothing in his non-verbal interaction to suggests that the reference should be to a physical manifestation of or to an index to such a course),[4] Hans treats the 'green driving' course as something that has already been mentioned. Consequently, the reference ties back to something that was previously made relevant in the session, more specifically to a point not long before, when the moderator mentioned that the theme month included the opportunity of winning a 'green driving' course. By mentioning this opportunity, the moderator challenged the ongoing accomplishment of the month as being both all about *public* transportation and, by in situ implication, of relevance only to commuters. Interestingly, at the time, this was accepted only by Jenni and Svend and the subject abandoned. Hence, at the time, the participants did *not* treat the course as a relevant challenge in their ongoing accomplishment of the month as being irrelevant to *their* specific incumbencies in the 'stage of life' device. In contrast, they *do* treat it as a relevant challenge when Hans reintroduced the subtopic.[5]

However, before continuing with Hans's work in line thirteen, I would like to go back a few lines and briefly mention Hans's work in line six. As it is, already before Birgit accomplishes the logic that 'if you are retired, you cannot be expected to have noticed the transportation theme month', Hans self-selects and initiates a turn. His turn overlaps in a non-transition relevant place, but, acknowledging the physical setting and the non-verbal interaction, it seems reasonable to suggest that Hans is not aware that Birgit has already initiated another turn. Birgit and Hans sit diagonally opposite each other, and Hans is oriented downwards towards his cake when Birgit addresses the moderator and starts her turn in line three. Furthermore, Hans readily breaks off, thereby presenting his overlap as a 'mistake', not as an attempt to interrupt Birgit's work in progress. Still, precisely

4 Furthermore, as shown below, line 13 comes off as *repair* of Hans's overlap in line 6, *not* as anaphor to something made known there.

5 The past tense 'det *bed* jeg også mærke i'/'that *did* I also take special notice of' makes the reference ambiguous concerning whether Hans noticed the course *also* during the theme month. The past tense could suggest that Hans noticed the course as well at that time; that is, before this meeting and the moderator's mention. However, what can be unambiguously known from data [from the enclosed excerpt] is that the moderator has mentioned the course earlier, and that Hans at that point acknowledged her mentioning with a 'yes'. Further, it does not really matter for the presently established argument whether or not the past tense adds a further layer of reference to the utterance.

because Hans does not recognise that Birgit wants to continue talking, his attempt can be heard as being unaware that Birgit's second pair part in lines three, five, seven and eight is otherwise treated as *clearly* following from her preface in line one. In other words, Hans's 'mistake' reveals that he is not aligned with the otherwise common logic of the situation. The ensuing interaction supports this understanding. By challenging the otherwise cooperatively accomplished logic that retired people should not be held accountable for the theme month, Hans challenges the notion that the bound predicate 'retired' is the only predicate that matters with regard to the theme month topic.

By repeating the two word reference 'det der' ('the') from line six, Hans's first turn constructing unit ('det der kør grønt kursus'/'the green driving course') comes off as self-repair. In this way, the unit carries on the project of the overlap, and as such, it tentatively supports the suggested reading of line six. Relatedly, the turn in lines thirteen to sixteen is displayed as being yet another possible second pair part in the overall topic talk project. Here, Hans's utterance 'det bed jeg også mærke i'/'that did i also take special notice of' should be noted. So far, none of the other participants have taken *special* notice of anything. For instance, recalling Jenni and Svend's joint venture in the beginning of Excerpt 20 (discussed in the above analysis), the participants have until this point worked pretty hard to remember anything at all. However, recognising Hans's relatively late intro in the multiple second pair part sequence, his unit comes off as a sequential, expected upgrading (others have after all noticed something). Accordingly, it would be reasonable to hear his unit primarily as displaying the 'series-ness' of the multiple second pair part sequence. For instance, 'også'/'also' comes off as displaying Hans's turn as just one more in a series, and 'bed mærke i'/'took special notice of' constitutes the expected 'upgrade' in a sequence oriented to what the participants may have noticed. Furthermore, Hans's wording ties to the work in progress, partially reformulating the vocabulary employed by Birgit in line ten.

Consequently, Hans displays his mention of the 'green driving' course as something that co-constitutes the common project; that is, as yet another second pair part. In remarkable opposition to the already delivered second pair parts, Hans's second pair part comes off as a *preferred* response. By calling for resources to sustain the topic (the interest in the 'green driving' course), Hans encourages the moderator's topic-proffering first pair part (in combination with her prompt mention of the green driving course). Accordingly, his response comes off as preferred.

Recalling the systematic reversion of the ordinary differential expansion relevance of preferred and dispreferred seconds in topic-proffering sequences (Schegloff, 2007a: 169), the remaining part of Hans's turn displays an expected piece of justifying, elaborating post-expansion. From the middle of line fourteen to line sixteen, Hans substantiates and accounts for his base second pair part, as he classifies himself as someone who would like to learn green driving. Literally, he merely says, 'fordi det kunne jeg godt tænke mig at lære'/'because I'd like to learn that', but since he uses the utterance to qualify his reopening of the subtopic, his wish to learn comes off as related to the topic concerned, namely the 'green driving' course. This understanding is confirmed in Hans's next turn in lines nineteen and twenty. However, before looking into these lines, note the modality in the predication. The unbinding and slightly subjunctive modality in the verb construction (the combination of the auxiliary verb 'kunne'/'would' and the infinite verb 'tænke [mig]'/'like') presents the predicate as a *possible personal wish*; that is, as a wish one ('voluntarily') can or cannot personally hope or aim to fulfil. Importantly, Hans's account thereby comes off as 'safe' for him and 'unthreatening' for the others. Since the account draws on what Pomerantz (1980) refers to as a "type 1 knowable",[6] a highly personal piece of information, Hans not only has every right to draw on it (accordingly, it is 'safe'), he also has the full responsibility for doing so (accordingly, it is 'unthreatening').

In lines nineteen and twenty, Hans provides one more account, further substantiating his base second pair part. The premise marker 'for'/'because' (line 19) presents the following as an explanation for his wish. The explanation, then, is Hans's reference to the fact that he has 'aldrig nogensinde'/'never' been a very good 'chauffør'/'chauffeur'. This explanation comprises an interesting piece of category work. Firstly, the reference to a long time span invokes the already established 'elderly' category in the 'stage of life' device.[7] Whereas *time* may or may not be a relevant predicate in the 'stage of life' device, to see that it is Sacks's economy-rule should merely be extended to devices. Since the timespan-related

---

6 Pomerantz (1980) distinguishes between two forms of knowledge when she observes that, "Type 1 knowables are those that subject-actors as subject-actors have rights and obligations to know", whereas "Type 2 knowables are those that subject-actors are assumed to have access to by virtue of the knowings being occasioned" (187). When Stivers, Mondada and Steensig (2011) write about the concept of "epistemic responsibility", they point out that according to Pomerantz' distinction, one has a right as well as an obligation to know certain things, e.g. one's name, what one is doing, and how one is feeling. However, they importantly argue that epistemic responsibility extends such highly personal information, so the point is rather that type 1 knowables are most often treated as related to a higher degree of responsibility than type 2 knowables (17–8).

7 Of course, the upgraded modality (ever never) also reveals that the account is the second account given.

'stage of life' device is already made relevant in the preceding interaction, it is *economical* to assume the relevancy of that very same device, now that time is once again made relevant. Then, drawing once again on the economy rule it can be seen that it is more specifically the 'elderly' category that is co-invoked. Since it is already established that the participants belong to the 'stage of life' device, due to their incumbencies of an 'elderly' category, Hans's reference to past times does, of course, come off as a reference to his membership of this category. Secondly, by resisting the assessment 'a good driver', Hans invokes the predicate 'driver' as being bound to the category 'elderly' (and, in fact, also to previously upheld memberships within the 'stage of life' device). Since the appraising adjective 'good' comes off as inextricably bound to the predicate 'driver', Hans's work necessarily presupposes that the predicate 'driver' does already apply to him (and to other memberships within the device).

The category work fulfils the challenge Hans projected by delivering a preferred response. Hans suggests that with regard to his incumbency, the category 'elderly' is not only characterised by the predicate 'retired', but also by a wish to learn green driving. Thereby rendering the theme month relevant, Hans challenges the otherwise accomplished isolation. However, he ascribes the new predicate only to himself. Consequently, 'green driving' is displayed as a voluntary aspect of driving; you can consider learning it if you feel like it, but the phenomenon is not in any way treated as something car drivers should or ought to consider.

Hence, Hans's categorisation work does not specifically entail moral evaluation or establishment of categories characterised by the driver predicate. However, even though the other participants to the setting may not be expected to treat Hans's category work as problematically negotiating their co-memberships of the 'elderly' category, they may be expected to treat it as problematic that Hans does at all make the 'driver' predicate categorially relevant for the practical purposes of delivering a topic-encouraging second pair part oriented to a topic that has otherwise been co-constituted as not categorially relevant for the parties to the setting.

As already mentioned and as examined in the proceeding, the other participants do consider Hans's work in need of (re)negotiation. Furthermore, they do so in a remarkable way, as they co-construct a more specified version of the 'elderly' category, namely a version that is already bound by a 'green driving' predicate, which renders it irrelevant for members of the category to even consider taking a 'green driving' course. As we will see shortly, this interaction comes off in an especially rich and layered post-extension. In the following, I firstly demonstrate how Carin and Svend dismiss Hans's interest in the 'green driving' course. Since the dismissal is performed in a complex part of the sequence, the analysis contains some long-winded observations on structure in the aim to determine what is in fact going on and for what practical purposes. Secondly, I show how Hans elaborates and defends his second pair part against the others' rejection of it.

The interaction that follows Hans's second account includes what is treated as a subordinate orientation towards this account as self-deprecation. In lines twenty-three to twenty-seven, Hans reveals that normatively speaking, his act in nineteen to twenty was an act of self-deprecation that calls for disagreement as its preferred response (Heritage, 1984a: 1969). Accordingly, he can be heard as pursuing a preferred response, but the pursuit is carefully marked as of *structural* relevance only. Firstly, before line twenty-three, no one has taken the opportunity to treat Hans's action as self-deprecation. For instance, no one self-selects in the pause in line twenty-one, fulfilling the deprecation by countering it. On the contrary, the moderator displayed agreement with Hans's self-characterisation by shaking her head slightly in line twenty-two. Then, by underlining that he would himself like to counter the self-deprecation (lines twenty-three and twenty-four) and that others are welcome to do so (twenty-four to twenty-seven), in lines twenty-three to twenty-seven Hans explains that the situation might pose a structural problem. However, since he *himself* does the explaining, and since the moderator and Svend laugh as a response (in lines twenty-eight to twenty-nine), the common orientation to self-deprecation reveals that the latent structural problem is exactly only a *structural* problem, not an otherwise serious or relevant problem for the practical purposes at hand.

Against this, Carin's turn in line thirty-two comes off as a notably delayed response to Hans's work in lines nineteen to twenty. For instance, she could have self-selected in the pause in line twenty-one, but she does not respond until *after* Hans has pointed to the latent structural problem. Also, she does not respond immediately after Hans has addressed the matter of self-deprecation; rather, her response is further 'delayed' until after the laughing. The delay obviously represents dispreference, but it is by no means easy to establish the overall preference orientation of her work. Her response may or may not be part of the humorous treatment of Hans's self-deprecation. However, as the sequence progresses, Carin's work is more unambiguously treated as being primarily a part of a dispreferred challenge of Hans's work.

This complexity is not reflected on the categorial level. Accordingly, by saying 'årh men du kører ganske udmærket'/'oh but you drive pretty decently' (lines thirty-one to thirty-two), Carin challenges Hans's attempt to resist a 'good driver' predicate. Her challenge is presented via the disagreement displaying preface 'årh'/'oh' and via the disjunctive connective 'men'/'but'. The intertwined 'membershipping' takes place when she categorises Hans as someone who drives decently. Subsequently, Carin and Hans each initiate a turn in lines thirty-six and thirty-seven, respectively.[8] Consequently, Carin breaks off, but in line thirty-eight, she continues, self-repairing. She merely says, 'og ikke for stærkt'/'and not too fast', but since it comes off as being linked to her 'membershipping' in lines thirty-one and thirty-two, it displays the compound suggestion that Hans is characterised by the predicate 'drives rather decently and not too fast'. However, it remains difficult to determine what Carin wants to do with this predication of Hans. In particular, it is difficult to determine whether Carin's category work is a preferred response which aims at rejecting Hans's self-deprecation only or a dispreferred response which aims at rejecting this self-deprecation *only* because it is embedded in the base second pair part. However, the ensuing interaction steers the purpose of the category work and, by implication, the interaction as such in a rather specific direction.

In line thirty-nine, Svend orients his body towards Hans and self-selects at the transition relevant point displayed at the end of Carin's clarification of the 'driver' predicate. Furthermore, Svend delivers a refined piece of category work, subordinating Hans's category work. More specifically, Svend negotiates the 'elderly' category in a way that renders Hans's work unthreatening for the ongoing topic isolation work. In this way, as I show shortly, Svend's category work in lines thirty-nine and forty comes off as disaligned with the stance displayed in Hans's work and as co-constituting a dispreferred response to Hans's second pair part to the moderator's first. Hence, Svend's work orients to Hans's potential self-deprecation not as self-deprecation, but as a part of Hans's second pair part. Or, in other words, by disagreeing with Hans's self-deprecation *and* dismissing it as a topicalising second pair part, Svend does two things at once. Consequently, considering the *entire* excerpt, Hans's account in lines nineteen to twenty is *primarily* accomplished as a structurally expected elaboration of Hans's base second pair part (lines thirteen to fourteen) and his first account (lines fourteen to

---

8 Since he is oriented towards the moderator (note his gaze), it does not come off as an intentional interruption. Rather, it comes off as not really being oriented towards Carin, but, instead, to the moderator's laughter. Thereby, it sustains the humorous treatment of the formal structural problem invoked by the 'missing' (normatively speaking) rejection of his self-deprecation. Furthermore, his 'til en afveksling'/'for a change' is readily heard as oriented to whether anybody 'for a change' would reject (or have rejected) his self-deprecation. That is, the line comes off as a reached for exaggeration of the structural problem that as such emphasises the humorous treatment.

sixteen), *not* as an isolated instance of self-deprecation. As such, it makes sense to approach not only Svend's, but also, retrospectively, Carin's responses to Hans as being the type of post-expansion Schegloff refers to as "rejecting/challenging/disagreeing with the second pair part" (Schegloff, 2007a: 159).

Against these sequential observations, it is time to consider the co-constituting category work more thoroughly. By referring to Hans's age ('i din alder blom der er du grøn chauffør'/'at your age blom you are a green driver'), Svend invokes the 'stage of life' device and the 'elderly' category (once again according to Sacks's economy rule). However, whereas Hans invoked the 'elderly' category in relation to the predicate 'driver', Svend invokes it in relation to the more specific predicate 'green driver'. Consequently, being or not being a green driver is presented as a question of age, not a question of green driving courses, for instance. This evidently challenges Hans's self-ascription to a membership of drivers who want to *learn*, since it pre-empts any such wish; you cannot learn to do something you already know how to do.

Since the 'elderly' category is accomplished as a collectivity category, Svend's work not only ascribes to Hans's work; rather, when it is ascribed to Hans, it is simultaneously ascribed to the entire collectivity. Hence, if Svend succeeds in rejecting Hans's claimed topic-embracing resources (his wish to learn), his work will automatically isolate him and the other participants from this possible threatening topic extension. The entire 'elderly' collectivity, it must be assumed, is already characterised by a 'green driving' predicate that pre-empts any reasons for considering a green driving course.

As mentioned, Svend's work negotiates the purpose of Carin's preceding work, displaying it as co-accomplishing the pre-empting of Hans's topic embracement. In line forty-one, Carin responds to Svend's work. Even though her response is constructed as minimal, it is not delayed and it clearly displays alignment with the polarity of Svend's turn. Furthermore, as I demonstrate in a while, it should be noted that Svend's and Carin's ensuing interaction, among other things, displays a rather specific understanding of Carin's already delivered work.

In line forty-two, Svend, as expected (normatively speaking), qualifies his challenge of Hans's second pair part. In particular, Svend points to how Hans (and other incumbents) is to be understood as being a green driver. He exemplifies the local meaning of the predicate when he says, 'du ræser jo ikke af sted'/'you know you're not rushing ahead'. The line is audibly undelayed, its onset almost coinciding with that of Carin's brief response. Furthermore, it ties back to his former work through the explanatory adverb 'jo'/'you know'. Accordingly, the 'green driver' predicate comes off as being related to Hans's speed dispositions. The supposed 'speed disposition' is formulated negatively, as a matter of Hans *not* rushing ahead.

On a more specific level, the negative formulation ties back to lines thirty-six and thirty-eight in which Carin substantiated her assessment of Hans as being a rather decent driver who does not drive too fast. Since Svend's account is also formulated negatively, it displays affiliation with Carin's work. Particularly, Svend's formulation reflects on Carin's work as part of the project he also contributes to, namely the project of challenging Hans's second pair part. On a more general level, the negative formulation embeds and, to a certain extent, invokes the *positive* formulation 'driving too fast' or 'rushing ahead'. As argued, the negative formulation exemplifies a predicate bound by the category 'elderly' within the device 'stage of life', but at the same time Sacks's consistency rule leads us to hear the embedded *positive* formulation. Furthermore, the positive formulation invokes a predicate bound by different category incumbencies within that same device. As shown in the preceding part, the 'retired' predicate not only isolated the participants to the focus group setting from any responsibilities of knowing, it also (more and less implicitly) invoked such responsibility of members bound by *non*-retired predicates. In the exact same way, the negative description 'does not drive too fast' not only extends the topic isolation, it also invokes other categories that are *not* isolated and *not* non-responsible.

As it is, Carin accepts Svend's work, almost whispering 'yes' in line forty-three. Then a lengthy pause follows, before Hans acknowledges Svend's focus on him. This happens as Hans initiates other-initiated repair, revealing that he does not understand what Svend is trying to get at (line forty-five). Svend performs other-initiated self-repair, repeating and elaborating on his 'you're not rushing ahead' account (lines forty-six and forty-eight), and Carin supports this with brief, agreeing responses (lines forty-seven, forty-nine and fifty-one). Thus, they cooperatively display the observed problem at hand as a matter of lacking explanation, and as Hans shows alignment through his 'yes' in line fifty-three, the post-expansion seems finally to be at a closure relevant point. This is also evident when Svend in line fifty-four sums up their accomplished common understanding: since Hans does not rush on but drives quietly and easily, his driving is probably as green as possible. The scope of this action should be noted. Even though Svend is oriented to Hans, it is, it should be recalled, an orientation that treats Hans as an 'exemplar' of the collectivised category 'elderly'. Hence, significantly, Svend's account comes off as applicable to all the participants.

However, closure does *not* follow. Rather, in lines fifty-six to sixty, Hans elaborates and defends his second pair part against rejection hereof. The defence comes off as a negotiation and not as a direct rejection of Svend's challenge. This is evident from a couple of features. It is prefaced by a display of alignment; it comprises different other dispreference mitigation devices; and it comes off as an *elaboration* (and not, for instance, as a repetition) of his challenged second pair part. Specifically, Hans does not directly resist Svend's attempt to categorise him as belonging to an 'elderly' category that binds a 'not rushing ahead' predicate. Regardless of this carefully performed negotiation work, Hans's defence is rejected immediately. For the sake of lucidity, I break this skilled negotiation work down into a number of steps.

Initially, Hans' response does not come off as dispreferred.[9] By saying 'jo det er rigtigt'/'yes that's right', Hans initiates the turn with a turn construction unit displaying aligned agreement. At this point, the agreement comes off as fulfilling Svend's 'elderly' category work, including its last-mentioned implication that its incumbents probably drive in as green a way as possible. However, Hans's next units subtly question the observably implied assumption that *speed* is all there is to consider in relation to the relevancy of green driving. Hence, Hans questions whether the green driving course (which he drew on as a resource for sustaining the proffered topic) is rendered irrelevant, simply because he does not drive too fast. Consequently, if Hans's turn is considered in full, its initial display of agreement perhaps comes off *predominantly* as a preface-mitigating dispreference in a skilfully negotiating defence. In more detail, Hans continues in line fifty-six, saying 'men altså'/'but well'. The disjunctive connective 'men' projects some degree of contrast to the aligned stance displayed until that point, and the hedging 'altså' carries on the projected dispreference. Furthermore, Hans opens line fifty-seven with yet another dispreference mitigation marker: a hesitating 'øh'/'um'. In this way, all things considered, the anticipated 'base defence' is projected as exactly a *defence*. As such it comes off as a dispreferred response.[10]

---

9 The response is delayed, but since Hans's responses overall are delayed (they are also pronounced more slowly than the other participants' responses) and since he, from time to time, displays hearing problems (e.g. in his turn in line forty-five), the lengthy pause in line fifty-five does not come off as a significant display of dispreference.

10 Schegloff (2007a) points out that second pair part speakers can respond to challenging/rejecting post-expansions in (at least) two different ways. They can either attempt to defend their challenged second or they can provide a new, alternative second pair part to the initiating first (161). Seeing as he treats this distinction as being interesting at the structural level of cooperation rather than at the affective level (Stivers, 2008), I, because of the display of structural dispreference, see Hans's turn in lines fifty-six to sixty as an action that defends his already provided second pair part, not as an action that offers an alternative.

Saying that he would like to drive *economically* (lines fifty-seven to sixty), Hans's 'base defence' comes off through an elaboration or explication of the project of the challenged second. By referring to the idea of economic driving through the closeness reference 'her'/'this' (in 'det her med'/'this thing about' in line fifty-seven) and the commonplace display 'og sådan'/'and so on' (line fifty-eight), Hans treats the idea of 'driving economically' as a shared idea. In this way, the 'defence argument' itself comes off as mitigating the dispreference of delivering it at all. More specifically, this happens as the argument merely appears to elaborate on Svend's already suggested category work with a known and acknowledged dimension. Nonetheless, since Hans' work displays a defence of his second pair part, structurally speaking, it treats 'driving economically' as a relevant predicate, considering the relevancy of the second (which means considering the relevancy of the 'green driving' course). Hence, it is challenging, even though it does not directly resist Svend's category work.

Considering Hans's account work, note that his 'base defence' is intertwined with yet another dispreference mitigation, namely an account of the relevancy of his response. Accordingly, by saying 'det vil jeg faktisk meget gerne'/'i'd actually really like that' (lines fifty-nine and sixty), Hans completes his turn by more or less repeating his first account (lines fourteen to sixteen). In this way, he once again draws on an account for which he can claim epistemic responsibility, with certainty, principally unthreatening. This time, the account is expectedly (structurally speaking) upgraded, in that 'faktisk'/'actually' strengthens the certainty of the feeling, whereas 'meget'/'really' strengthens the degree hereof.

Whereas people are generally treated as being fully responsible for their own inner life, the understanding of the shared world causing this or that feeling might of course be treated as highly negotiable. Hence, while Hans might have the epistemic right to, and be responsible for, claiming the experience of a specific wish, the understanding of the shared world causing this wish might be highly contestable, and in the case at hand it is. The present problem is a matter of whether or not the participants can agree on a world in which it makes *sense* that Hans wants to learn green driving and, more specifically, economic driving. As it is, the participants fail to agree. Notwithstanding Hans's hard mitigation work, his defence work is also challenged, and immediately so. In the first transition relevant place after Hans's closing account work, Carin self-selects and delivers an unmitigated dispreferred response that rejects the implications of Hans's account.

In line sixty-one, Carin says, 'men det har du da altid gjort'/'but surely you have always done so', thereby pre-empting Hans's account in much the same way as Svend's 'in your age' work pre-empted Hans's account of his original second pair part. Implying that Hans is *already* characterised by a 'driving economically' predicate, she renders it irrelevant for him to strive for it, and the epistemic adverb 'da'/'surely' underlines her orientation to Hans's work as a moral transgression of their already shared background knowledge. The pre-emption weakens the otherwise accomplished rights of the 'elderly' category, but, importantly, it similarly radicalises the primary isolation work. This should be considered in more detail.

By invoking the 'already-ness' of the predication by using the definite adverb 'altid'/'always', Carin presents an extension of the ongoing category work within the 'stage of life' device. As shown, this category work has so far focused on the accomplishment of the 'elderly' category as an account resource that could isolate the participants (as members of a specific collectivity) and thus prevent them from being held accountable for knowing about the transportation theme month. However, so far, the participants have drawn on the collectivity category not only to withdraw their collectivity from any responsibility, but also to place responsibility on another collectivity within the device, namely the 'middle age' collectivity (negatively invoked through the commonly self-ascribed resistance of the employee predicate). Yet, by rejecting Hans's defence attempt with a reference to what he has *always* done, the rejection does not seem to likewise elaborate the 'elderly' category by contrasting it to other categories within the device. Rather, it appears to elaborate the device all over. Acknowledging Sacks's economy rule, it would not be sustained to suggest the relevancy of any new device. Accordingly, the reasonable hearing of Carin's line is that the time adverb 'always' ascribes the predicate 'driving economically (and so on)' to all possible categories (that is, stages) within the 'stage of life' device. On the one hand, these implications of Carin's rejection 'loosen' the work on the 'elderly' category, but, on the other, they sustain the isolation. Since 'always' ascribes the 'driving economically' predicate to the entire device, it projects (some degree of) isolation against the 'green driving course' to all of it.[11] In this way, Carin's work nicely illustrates the 'for all practical purposes' character of the ongoing category work. Isolation is the objective, and the aim of the category work is to contribute to this process of isolation.

11 Structurally speaking, Carin's 'always' rejection comes off as an expected upgrading of Svend's 'already' rejection of Hans's original second pair part – in much the same way that Hans's account of the defence came off as an upgrading of his account of the second pair part. However, that observation does not change the mentioned implications of the category work co-constituting the rejection.

Carin's rejection work and the radicalised isolation work are successful. By providing a brief aligning response (line sixty-two), Svend fulfils it, and, importantly, Hans does not pursue the 'green driving course' topic further. However, as such the sequence does not end with Carin's rejection and Svend's fulfilment. Rather, it includes yet a post-expansion. This time, the expansion comes off as a form of topicalization,[12] and Schegloff (2007a) points out that "such expansions often become subject to the organizational contingencies of topic-talk" (158). Accordingly, it might be optimistic to ever expect norm-conforming closure. However, the expansion does, on its own 'organisational contingent topic-talk' premises, generate if not sequence closure, then sequence transition. In particular, this happens through a delicate reframing of the topic. I would like to provide a few comments on these remaining expansion lines, especially on Svend's category work on which the reframing is based.

Svend utilises Hans's defence in one more way: as a starting point for reframing the topic at hand. This act of reframing facilitates the opportunity to accomplish yet another isolation category. Svend draws on the practice of reframing to ascribe to a membership of persons who already hold all relevant knowledge on green/economic driving. Sequentially, the topicalization comprises lines sixty-four to 104. Svend initiates the expansion in line sixty-four, and the moderator initiates her accomplishment of the reframed topic in line 104 (a few lines later she introduces a specific topic proffer on the reframed topic).

Whereas Svend's brief response in line sixty-two showed structural alignment with Carin's rejection, his turn in lines sixty-four to sixty-seven reveals a certain empathy with Hans's defence. This occurs as Svend focuses his body on Hans, pats him on the arm and focuses verbally on his wish to drive economically. In this way displaying an evaluative stance that matches the one evident in Hans's account, the turn comes off as predominantly *affiliating* with Hans's defence work. As I demonstrate below, Hans draws on the affiliation in order to steer the sequential agenda in a direction that particularly heightens his agency.

Svend's empathy is displayed through a consequent that co-constitutes a conditional in which Hans's defence is the antecedent (lines sixty-four to sixty-seven). In this way, Svend treats Hans's turn as being a request for advice on economic driving, and Svend's turn comes off as the preferred response supplying this advice. Thereby, Svend negotiates the understanding of Hans's work which

12 Schegloff (2007a) stresses that the non-minimal type of post-expansion 'topicalization' is used to mark some utterance part as one of special interest and worthy of further on-topic talk (155).

was not readily heard as a direct request for advice, but rather as an account of Hans's interest in a green driving course. Hence, Svend's affiliation display co-constitutes a negotiation of the meaning of Hans's work. Furthermore, by co-constituting, Svend invokes a new categorisation for himself that transcends the categorisations projected by the moderator's topic proffer.

Svend suggests that if Hans wants to drive economically, he should consider buying a new car (lines sixty-four to sixty-five). Firstly, it should be noted that Svend easily invokes a category of car owners. 'Considering buying a new car' straightforwardly invokes the possessive predicate 'owns a car', which in turn invokes the category. Further, Svend invokes a rather interesting membership of persons who hold knowledge not only of cars as such, but of cars that encourage economic driving.[13] By saying 'så er det du skal have'/'then it is what you must have' (line sixty-four), the claim of 'holding knowledge' comes off through the confident modality. Furthermore, authoritarian ownership is underlined throughout the expansion in a number of features: for instance, when Svend displays a high degree of certainty in his own argument, saying 'det siger sig selv'/'it's evident' (line seventy-eight, when he displays a right to assess the 'aptitude' of Carin and Hans' current car (lines eighty-five, ninety-two and ninety-four), and when he displays the validity of his knowledge, providing a concluding 'personal experience' story (lines ninety-five to 102).

Consequently, Svend has projected a reframed topic where he is in charge. Whereas Hans's interest in the green driving course made the transportation theme month tentatively relevant, Svend (in cooperation with Carin and, remarkably, Hans, [for instance line eighty-nine]) has generated the accomplishment of a slightly different story: Hans is not interested in the theme month as such, but in economic and green driving. Since Svend holds all appropriate knowledge on this matter, the theme month is still rendered irrelevant. Hence, while the moderator (in the proffering first pair part) and the participants (in the different second pair parts) initially accomplished memberships bound by different predicates of ways and degrees of knowing about the transportation theme month, Svend here ascribes to some sort of 'consultant' category invoked by the bound predicate 'holds knowledge of green and/or economic driving'. Importantly, this predicate is treated completely differently than any 'holds knowledge of the transportation theme month' predicate. Particularly, whereas Svend was

13 Recall Sharrock's (1974) paper on 'owning knowledge'; particularly, recall that owning knowledge might not be that different from owning other objects, in that it might generate much the same types of rights, obligations and responsibilities (52).

cautious of being held accountable for knowing about the theme month, he is rather upfront when he displays himself as knowledgeable about green and/or economic driving.

In a way, all that has shifted is the ownership of expert status. However, in line with Rose's observation that in current governmental rationalities, individuals can best fulfil their political obligations in relation to the wealth, health and happiness of the nation, not when they are bound by relations of dependency and obligation, but when they seek to fulfil themselves as free individuals, it is exactly the re-claiming of ownership of knowledge that provides Svend with the interactionally accomplished power to fulfil himself as a *free* individual, not as a citizen stirred or bound by a patronising municipality.

The utilisation of autonomous knowledge

As mentioned, the purpose of this section is to show how the participants pre-empt the municipal attempt to conduct the citizens' transportation conduct. The section demonstrates how the participants consider the municipal attempt inconsequential, seeing as they already hold the relevant knowledge about the suggested conduct. However, the following analysis additionally shows that the conduct in question is firmly accomplished as a right, not as an obligation, thereby notably decreasing rational accountability. As such, the analysis not only shows how rational accountability is pre-empted, in that it is demonstrated that the participants already comply with the rationality due to their *autonomously* held knowledge, it also shows how the call for accountability is radically decreased, as it is demonstrated that this knowledge binds an autonomous right, but not an obligation to comply with the conduct.

The excerpt discussed in the following stems from a session carried out after the municipal theme month. The session concerned the participants' reactions to the municipal transportation theme month, focusing on how they related the strategy to their own practices. The interaction represented in Excerpt Five is taken thirty-two minutes into the session. At that point, the parties to the setting have stopped talking about their experiences of the month as such and of the theme day at the supermarket. Furthermore, they have been talking for quite some time about the 'green driving' issues pointed out in the flyer they have so far not seen. In the interaction represented, the parties to the focus group setting accomplish the municipal strategy in and by pre-empting rational accountability for the strategy. This happens as the participants accomplish their *already* possessed knowledge of economic driving as appropriate for all practical purposes. Furthermore, the participants accomplish economic driving as an individual choice; that is, as an individually achieved *right*, not as an *obligation*.

Excerpt Five

Participants: Moderator: mod. Julie: jul. Sussi: sus. Berit: ber. Tine: tin

```
1    tin/jul  ((looks at mod))

2    sus      ((looks towards tin))

3    ber      ((looks seemingly straight ahead, towards sus))

4             (.)

5    mod      ((looks seemingly across, towards sus))
              øhm:
              um

6    sus      ((looks at mod))

7    mod      du sagde at øh <speederen=
              you said that um the accelerator=

8    tin      ((looks at sus))

9             =måske var bedre på: (.)
              =was maybe better at (.)

10            ((looks at sus))

11   sus      hundredetyv(h)e=    [=he-he-he
              hundred and twenty= [=he-he-he

12   mod                          [end hundredeti
                                  [than hundred and ten

13   jul                          [((looks at sus))

14   sus      [((laughs))

15   tin      [((still looking at sus, laughs))

16   jul      [((still looking at sus, smiles))

17   mod      ((looks down towards papers))

18            ..h da ifølge ifølge det materiale
              ..hh when according according to the material

19            ((looks up))

20            mette har:
              mette has

21            ((looks down again towards papers))

22            gi:'d- har- har lavet dar
              given- has- has made there's
```

23 øhm (.8) da øh v- ka en (1.5) en
um (.8) then um can an (1.5) an

24 >forøgelse af hastigheden< fra
increase of speed from

25 hundredeti til hundredetredve (.).hh ka
hundred and ten to hundred and thirty (.) .hh can

26 ((looks up))

27 gi en stigning af forbruget på op til tyve procent.
increase consumption with as much as twenty percent

28 (1.1) er det: >al'så på grund af- (.) af
(1.1) is that >that is because of (.) of

29 vindmodstanden (.6) er der nogen af jer dar (1.2)
wind resistance (.6) is any of you (1.2)

30 dar tænker over (.9) a'så er det-
considering (.9) well is it

31 ((tin turns back towards the moderator after placing a hanky in the pocket of her jacket hanging on her chair))

32 er det blandt andet så'n nogen ting du
is it for example things like that you're

33 tænker på når du sætter fartpiloten på motorvejen
considering when you're seting the speed pilot at the motorway

34 (.7)

35 tin ((smacks lips))

36 (1.1) ja hvis jeg pl- (til å) sætter den tit på
(1.1) yes if i (used to) often setting it to

37 hundredetyve fordi så synes jeg man har en
hundred and twenty because then i think you have a

38 god fart hv- hvor- hvor ikk' det-
good speed wh- where- where it doesn't

39 ((makes movement with her hands))

40 kører for vildt
drives too wildly

41 mod ja
yes

42 (1)

43 tin øh å osse hvis dar meget trafik
um and also if there's a lot of traffic

44 mod ja
yes

45 tin =så synes jeg nogengange hundredetredve
then i sometimes think that hundred and thirty

46 næsten er for stærkt (øhm) (2.1) å det
almost is too fast (um) (2.1) and that

47 har da- jeg har da gjort det nogengange
has of course- i have of course done it sometimes

48 fordi at je- je- jeg ved det er billigere.
because i- i- i know it is cheaper

49 mod ja
yes

50 mod [men der ikk'
[but there's not

51 tin [( )

52 det har- har min søn osse prøvet han
my son has has also tested it he

53 har lavet han har- hvor han kørte
has done he has- where he drove

54 hundredetyve hele vejen
hundred and twenty all way

55 mod ja
yes

56 tin =da kunne han se at han brugte mindre
then he could tell that he used less

57 han havde lavet så'n en eller anden test eller.
he had made such some sort of test or

58 mod ja
yes

59 (1.5)

60 tin .hh men a'så ikk' noget jeg så'n: (.)
.hh but well not like something i [adj.] (.)

61 hver gang. overhovedet
every time at all

62 mod nej
no

63 tin det kommer an på igen hvad man skal å
it again depends on what you have to do and

64		°å° (.7) .hh å hvad man skal nå om man and (.7) .hh and what you have to catch up with whether you
65		har god tid til å køre ogsåvidere ikk' a'så h.. have plenty of time for driving you know and so on right h..
66	mod	ja yes
67		(.6) (.6)
68	tin	a'så så'n som- som (.) på torsdag hvor jeg for example like- like (.) this thursday where i
69		ska nå en færge kvart over seks så skal jeg have to catch a ferry a quarter past six then i have to
70		videre til nyborg å- å være til tiden (.) .hhh da continue to nyborg and- and be on time (.) .hhh then
71		tænker jeg arh men jeg sætter den bare på hundrede i do think well but i just set it at hundred
72		((crosses arms, leans back))
73		fordi a'så så because well then
74		((gesticulates with hands))
75		kommer jeg måske for sent ikk' i will maybe be late right
76	mod	ja yes
77		(.)
78	tin	hvis man bare overholder if you just keep to
79		(.5) (.5)
80	mod	ja yes
81	tin	=a'så man ka ikk' altid (.) well you can't always (.)
82		((makes rolling movement with hand))
83		bare tænke (.9) økonomisk man er os nødt til just think (.9) economically you also need to

84		nogengange å passe sometimes to mind
85		((moves both hands categorically downwards))
86		nogen ting man skal some things you have to do
87	mod	ja yes
88		(2.1)
89	tin	°tingene skal jo ligesom hænge sammen° °things after all have to sort of fit together°
90	mod	er det øh (.) du- du siger ja is that um (.) you- you're saying yes
91		((nods towards sus))

The session was initiated thirty-two minutes before the interaction represented in the excerpt. In brief, the moderator started asking whether and, if so, how the participants had noticed the transportation theme month in November. After talking about the fact that they had *not* noticed anything, the moderator pointed out that the month (not surprisingly, she said) focused on how one can drive in a more green way. She mentioned that the project manager, Mette, had divided the transportation theme into minor focus areas, and that the first of these was concerned with fuel consumption and efficient driving. Then she asked the participants whether the idea of economic driving made them think of anything. The question prompted a lot of talk. Firstly, Tine said that she often considered 'such things' when driving, for example using the speed pilot in her car. Secondly, prompted by the moderator, the other participants said that their cars did not have such 'advanced' technologies. Furthermore, Sussi said that she did consider the economic aspect of driving, but that other things had to be considered as well. One such thing was that her foot rests better on the accelerator when she drives one hundred and twenty kilometres per hour than when she drives one hundred and ten kilometres per hour. Sussi's foot account was followed by thirteen minutes of talk, prompted by the moderator and referring to different aspects of the focus areas of the theme month, such as how to brake economically and how and when to use air conditioning. The air conditioning talk ended thirty-two minutes into the session, and from that point the interaction is represented in the excerpt.

Structurally speaking, the excerpt displays a relatively straightforward part of a 'multi second pair parts' topic-proffering sequence. Lines five to sixteen come off as a pre-expansion[14] sequence to the topic-proffering sequence. Lines seventeen to thirty-three are projected as the base first pair part in the topic-proffering sequence, and, importantly, the last part of this proffer is launched by a yes/no type interrogative. Then, lines thirty-five to eighty-nine come off as Tine's rather extended second pair part made relevant by the moderator's first. In the different category work delivered in these lines, Tine meticulously points out her stance to the proffer. Particularly, she draws on the specific constraints and affordances projected by the interrogative format of the proffer; that is, not only is Tine's response oriented to the action type generated by the interrogative in question, it is also oriented to the matter of aligning with the polarity of the interrogative. Lastly, Tine's turn in line eighty-nine ends her second pair part, and turning towards the other participants, the moderator fulfils this projection in line ninety. Acknowledging that Tine's work is part of a multi second pair parts sequence, only Tine's part, not the sequence as such, ends with the moderator's orientation to the other participants. In the following, I go through the pre-extension, the proffer and Tine's second pair part, bearing the intertwined relationship of these parts in mind.

Considering the pre-extension and the base first pair part, I am primarily interested in pointing out how the moderator co-constructs the following negotiations, projecting a specific context for the following second pair part. In particular, I demonstrate how the moderator reaches for a preferred response by withdrawing from a potentially 'problematic' proffer, shifting her focus to another participant and broadening the scope of the topic in question.

Lines one to sixteen represent a pre-extension sequence. As Schegloff (2007a) points out, 'pre-ness' is a constituting matter of specific types of activities, not of order of occurrence: "The prefix 'pre-' is meant to capture the action being done, and not 'occurrence prior to something else'" (34). Even though Schegloff draws on observations from everyday conversation, and I study focus group interviews of a particular form, his observations are nevertheless relevant to the present investigation. Schegloff's point is that 'pre-ness' and 'pre-works' are presented

14 Schegloff (2007a) points out that 'pre-expansion' refers to the first point at which some sort of two part unit can be expanded, namely before its first part. He observes that while it might sound problematic, in an empirically grounded analysis, to understand something by reference to something that has not yet occurred, it is not, since a pre-sequence is only to be understood as such if the parties to the conversation approach it as a pre-sequence. Schegloff says, "The parties to pre-expansion exchanges display an orientation in them to a base adjacency pair which *may* subsequently develop" (28, emphasis added).

and used in many different ways, and I do not see any reason why there should be a rigid boundary between different forms of 'pre-work' in everyday conversation and the 'pre-work' that co-constitutes focus group settings (such as the moderator's 'pre-work' for asking questions, introducing new topics, facilitating etc.).[15] Accordingly, I consider lines one to sixteen to be instances of important 'pre-ness' work. Initially, I approach the lines without considering that they are subsequently developed into a base adjacency pair. Then, only on this basis, I approach the lines as instances of important interactional work in relation to the following sequence.

The pre-sequence itself embeds a pre-sequence, namely a generic, minimally constructed summons-answer sequence in lines five to seven (cf. Schegloff, 2007a: 48–53). However, to see how this embedded sequence is accomplished, the previous interaction needs to be considered. The previous interaction has unfolded as a number of multi second pair parts sequences, in which the moderator has delivered the first pair parts, and the participant have delivered the seconds, cooperatively co-constituting the setting as a focus group setting. Considering this as well as the fact that the immediately preceding topic of air conditioning has just ended, it is feasible to approach the situation represented in the first few lines of the excerpt as anticipating yet another first pair part projecting a topic-proffering sequence.

Against this, the moderator's 'øhm'/'um' in line five displays a summons for a candidate receiver of an ongoing first pair part. It is not possible to tell from the video recording of the setting whether the moderator is in fact gazing at Sussi while delivering this summoning 'um', but if she is, she is of course not simply summoning a *candidate* receiver, but rather summoning Sussi in particular. However, it does not really matter whether the 'um' is directed at *any* possible future second pair part provider or at Sussi in particular. Rather, what matters is that the work accomplished in lines six and seven (retrospectively or not) presents Sussi as the

15 Relatedly, Schegloff (2007a) points out that 'pre-ness' does not always come off as 'pre-ness' on the very surface. In regard to an utterance form such as 'are you doing anything on Thursday night', it is displayed already at the surface that it is aimed at doing the job of a first pair in a pre-sequence to an invitation, but this is not always the case. Schegloff exemplifies this with an example where 'we have a thermometer' is in fact a 'pre-offer' and where the 'pre-ness' comes off as an answer to a question like 'why that now'. That is because there are no grounds in this particular case for mentioning this ownership. Before this utterance, someone has declared an intention to purchase a thermometer, and it is exactly on this background that the utterance comes off as a first pair part of a pre-expansion (35).

summoned part, regardless of whether or not the moderator's gaze had already picked her out. Sussi's gaze at the moderator in line six shows, firstly, her understanding that the 'um' was directed at her and, secondly, that the moderator's use of the personal pronoun 'du'/'you' in combination with the reference to Sussi's earlier mention of the accelerator fulfils the understanding of Sussi as the summoned part.

Whereas attention is in this way secured, the moderator does not deliver her first pair part. In contrast, lines seven to sixteen come off as yet another pre-expansion: this time not a 'mere' generic summons-answer sequence, but some sort of type-specific *pre-proffering* sequence, re-invoking Sussi's previously established argument of the comfortable position of her foot on the accelerator. In the present situation, there is no readily at hand reason for the moderator to mention Sussi's argument again. Consequently, the fact that she does so invites an understanding of the mention as possibly implementing some other action, namely, what I call a pre-proffer. This points to an understanding of the moderator's mention as 'checking out' the likely fate of a potential future base.

However, before continuing with the relation of the presequence to the projected base sequence, note the second pair part (of the pre). The moderator leaves her utterance 'open' in line nine and it is closed in cooperation by Sussi and the moderator in lines eleven and twelve. This cooperation not only ends the first pair part, but also incorporates Sussi's preferred second pair part response. Lines nine to twelve display the moderator's paraphrase of Sussi's earlier account and Sussi's acceptance of this. Hence, Sussi's response in line eleven encourages the re-invoking, and in effect it serves as a 'go-ahead' response to the pre-sequence. The moderator and Sussi thus collaboratively work out, firstly, that Sussi gave an account and, secondly, that this account could be part of a topic that it would be all right to reopen or elaborate on.

Against this, consider Sussi's laughter in line eleven and her and Tine's joint laughter (Julie is also smiling) in lines fourteen and fifteen (and sixteen). The laugh presents Sussi's earlier account as somewhat funny. Sussi starts laughing already at the end of her utterance in line eleven, displaying her laughter as oriented specifically to her own utterance. Also, the directions of and shifts in Tine and Julie's gazes should be noted. Before the moderator initiates the sequence with her 'um' in line five, both Tine and Julie look at the moderator. However, in the middle of the moderator's paraphrase of Sussi's earlier account, Tine's gaze shifts to Sussi (line eight), co-assigning the moderator's talk to Sussi. In line twelve, Julie's gaze also shifts and she, too, focuses on Sussi. Furthermore, also through the joint laughter and smiles in lines fourteen to sixteen do Tine and Julie look at Sussi (Sussi still focuses on the moderator, though). Hence, their laughter is clearly directed at Sussi and her earlier work, not at the moderator's paraphrasing, for instance.

More specifically, their laughter does not come off as directed at the moderator's phrasing, but it is the moderator's work that makes it relevant to revisit an earlier, somewhat amusing account.

By reinvoking Sussi's earlier account, the moderator can be heard as vaguely problematizing it. This is partly observable in the sequentiality. Simply by mentioning the account as part of a pre-extension, the moderator 'problematizes' it, treating it as being somehow still up for debate. Also, the modal adverb 'måske'/'maybe' in line nine furthers the display that the account is still up for negotiation. Accordingly, Sussi's preferred second pair part in effect accepts that her earlier account is not, for all practical purposes, satisfying. Against this, the laughter can be understood as being part of a common treatment of Sussi's account as not quite exhaustive. Hence, when the moderator continues in line eighteen it is on an observable, common understanding that Sussi's earlier account is not quite good enough.

The moderator initiates the delivery of her base first pair part in line eighteen (and non-verbally in line seventeen). In lines eighteen to twenty-nine, she mentions that according to the theme month material, fuel consumption increases when the speed increases, thereby introducing the projected topic (moreover, she adds that the relationship between speed and fuel is a result of increased wind resistance). Then, in lines twenty-nine and thirty, she asks whether anybody 'is considering', thereby accomplishing two aspects. Firstly, the suggested topic comes off as a matter of the relationship between fuel consumption and speed. Secondly, the relationship between Sussi's account and the 'new' topic is presented as the relationship between different accounts of one's speed. Remarkably, however, the moderator breaks off in the middle of line thirty, and from the second half of line thirty to line thirty-three, she shifts her focus, asking *Tine* in particular whether these may be the issues she contemplates when she uses the car's speed pilot. Hence, in the end, the moderator does not pursue Sussi's earlier problematized account.

In greater detail, the topic proffer includes a fair amount of self-initiated self-repair and notably long pauses. In the first, more introducing part of the proffer, the moderator repairs repeatedly in lines twenty-two and twenty-three; the latter includes two remarkably long pauses. However, non-verbally, the moderator is orientated towards her papers in lines seventeen and twenty-one, and she speaks with discernibly more speed, using a 'reading' tone of voice in line twenty-four. Considering this, the repairs and pauses come off primarily as an effect of the moderator's 'double' orientation to the participants and her papers, respectively. Hence, the moderator's self-repair and pauses come about as a matter of reading statements from the material in front of her aloud while simultaneously focusing on the participants (lines nineteen and twenty-six). On the other hand, the pauses, and repairs in the particularly questioning part of the proffer (lines twenty-nine to thirty-three) come off somewhat differently, showing that the moderator to some extent considers her own proposal problematic. This display of 'problematizing' is emphasised by the fact that the moderator notably shifts her focus to the participants. Whereas she initiated the proffer with a focus on Sussi and Sussi's earlier account, she finishes it with exclusive focus on Tine.

In lines thirty to thirty-three, the moderator projects Tine as next speaker. When the moderator says 'altså er det'/'well is it' in line thirty, Tine is facing backwards, focusing her attention on the pocket of her jacket that is hanging on her chair. Hence, there is nothing to indicate whether Tine is listening to the moderator talking. Since Tine's bodily non-orientation is observably visible from the moderator's position, the moderator shows her orientation to Tine specifically by simply repeating the line that was broken off (line thirty) in line thirty-two, when Tine is again oriented towards the moderator. Furthermore, this next speaker projection is evident from the moderator's stress on her reference to the speed pilot in line thirty-three. Not only does the moderator say the words 'speed pilot' which Tine talked about earlier, she also shifts to the unambiguously singular Danish pronoun 'du' (unfortunately, both the plural pronoun 'jer', which the moderator uses for instance in line twenty-nine, and the singular 'du', which she uses in line thirty-two, translate into the English 'you'; hence, the English version contains an ambiguity not present in the Danish one).

Besides projecting Tine as next speaker, the moderator also broadens the scope of her proffer. Whereas the moderator launches her concluding topic proposal through a somewhat constraining yes/no type interrogative, she carefully designs it to permit alignment and to presuppose as little as possible. In lines thirty-two and thirty-three, she says, 'er det blandt andet sådan nogen ting du tænker på'/'is it for example things like that you're considering'. Whereas the interrogative affords the moderator's explicit proffering and, by implication, her important projection of sequence development, it should be noted how carefully she reaches for a type-conforming and aligned response. Recalling Raymond's (2003) observations concerning the moral implications of interrogatives, it should be noted that the moderator's 'blandt andet sådan nogen ting'/'for example things like that' (rather than just a fuel-speed relationship) comes off as an orientation towards a type-conforming answer. Hence, reconsidering the moderator's orientation to Tine, it co-constitutes the introduction of the topic proffer in a way that does not imply too specific moral preferences. In particular, by focusing on Tine, the moderator utilises Tine's earlier mention of her speed pilot as a further assurance that she is heading towards a preferred response.

In summary, the moderator's work displays a thoroughly worked out orientation, not only towards a preferred response, but also towards a morally 'neutral' type-conforming response. By suggesting that economic driving might be related to Tine's above-mentioned speed pilot practice, the moderator is not just reaching for a preferred response; she is also projecting the idea that economic driving may be equivalent to current driving practices.

In lines thirty-five to eighty-nine, Tine delivers a rather extended second pair part. In these lines, Tine meticulously negotiates the topic via subtle displays of alignment, drawing on the specific constraints and affordances projected by the interrogative format of the proffer. Hence, on an overall level the second pair part can be said to comprise two parts: one predominantly oriented towards calibrating the type-conforming alignment, the other predominantly oriented towards calibrating the action-type alignment. The latter further comprises two parts: the first oriented towards accomplishing an autonomous right to drive economically, the second oriented towards accomplishing a *non*-obligation to practice such driving. Accordingly, the analysis of this complex second pair part comprises three foci: one on the more type-conforming oriented part of the response and two on the more action type-oriented part of it.

Firstly, Tine's verbal response in line thirty-six is initiated with a type-conforming positive response token ('ja'/'yes'), displaying alignment with the polarity of the yes/no interrogative it is, on one level, answering. Following this, Tine provides a double evaluation of her experience of speed, thereby taking the (projected) opportunity not only to provide a preferred response, but also to accomplish that she is in charge of her own conduct. Glossing, in lines thirty-six to forty-six, Tine points out that she uses the speed pilot more often when she is going a hundred and twenty kilometres per hour than when she is going a hundred and thirty kilometres per hour. The reason she gives for this is that she believes that a hundred and twenty is the "better speed".[16] The evaluation comprises two parts. In lines thirty-six to forty, Tines says that she often sets the pilot to a hundred and twenty [because she '*synes*'/'*thinks*' this is a good speed]. After

16 I do not treat it as important that Tine seems oriented towards a hundred and twenty kilometres per hour rather than a hundred and ten. I consider the issue of legal speed limits only to the extent that the participants are observably oriented towards such matters. However, it can be noted that the legal speed limit on Danish motorways is on some stretches a hundred and thirty kilometres per hour and a hundred and ten on others. Hence, legally speaking, it is not sustained to set the speed pilot neither to a hundred and thirty nor to a hundred and twenty, seeing as this speed would, by implication, be too high for the one hundred and ten stretches.

the moderator's 'yes' in line forty-one and a pause in line forty-two, Tine continues in line forty-three, adding that when there is a lot of traffic she sometimes '*synes*'/'*thinks*' that a hundred and thirty is almost too fast.

Tine's evaluation is interesting in two intertwined ways: firstly in a predominantly sequential way and, secondly, in a predominantly categorial way. Sequentially, the account is delivered just after the briefly delivered, but aligned base second pair part. Further, it is not oriented directly to the 'core' of the moderator's topic proffer. Particularly, the evaluation is not directly oriented to whether Tine finds the fuel/speed argument of the municipal material relevant or not. However, recalling how the moderator broadened the scope in her final question design, the evaluation comes off as aligned with such a broader understanding of the proffered topic, taking the provided opportunity to negotiate the proffered topic in a *preferred* response design. Responding with a 'yes' in lines forty-one and forty-four, the moderator fulfils Tine's answer as a preferred response.[17] In this way, Tine and the moderator cooperatively focus on the matter at hand in the interest of demonstrating that the participants already comply with the rationality in the suggested practice; *not* because it is suggested, but because it is *the* rationality.

Considering the category-work in Tine's evaluation, observe that in explaining or justifying her speed, Tine draws on a personal assessment of speed. By drawing on an evaluation of 'good speed' as her main, or, more specifically, firstly delivered reason for driving at a specific speed, Tine displays epistemic primacy. In more detail, drawing on this specific evaluation, Tine claims private 'ownership' of the point of view that causes her actions. Whereas speed could be a matter of law or, as Sussi has suggested, a matter of foot comfort or, as the moderator has suggested, a matter of fuel consumption, Tine makes it a matter of her autonomous experience of 'a good, not too wild speed' (lines thirty-eight and thirty-nine).

The display of epistemic primacy is also observable in the second part of Tine's evaluation. Reporting that she drives a hundred and twenty kilometres per hour rather than a hundred and thirty because she finds a hundred and thirty almost too fast, Tine stresses the factuality/identity verb 'er'/'is' with clear upward intonation,

---

17 Schegloff (2007a) points out that the preference for acceptance of a proffered topic, among other things, is underlined in the fact that most often, speakers of topic proffers will go for a second try if the first is not successful (173). It might be sustained to deduce backwards from this point and argue that not only are the moderator's yeses ordinarily seen as positive response tokens, they are also, the first in particular, delivered in places where something else, such as for example a second try, could have been launched.

thereby substantiating the importance of her own primacy and authority; of her own evaluation of reality. Consequently, Tine is drawing on her own autonomy which entails a *right* to drive at a specific speed; or, in other words, she is appropriating her way of driving, simultaneously drawing on and maintaining a 'freedom as autonomy' doctrine.

Tine's further work subtly negotiates the overall display of alignment in her second pair part. On the one hand, Tine establishes that her knowledge-possessing membership entails a right to drive economically, thereby at one and the same time weakening the type-conforming alignment and increasing action type alignment. On the other hand, Tine shows that she does not ascribe to any membership which entails an obligation to drive in such a way, thereby also definitely weakening action type alignment.

It is difficult to determine exactly where the second part of Tine's second pair part begins, since it is difficult to determine if the last part of line forty-six and the first part of line forty-seven are treated mainly as an interrupted attempt to continue the first part or as the beginning of a new attempt. Regardless, in lines forty-seven and forty-eight Tine says, 'jeg har da gjort det nogengange fordi at jeg ved det er billigere'/'i have of course done it sometimes because i know it is cheaper', and it readily comes off as a re-orientation to the moderator's proffer. Saying 'jeg har da gjort det nogengange'/'i have of course done it sometimes', Tine can be heard as rephrasing her type-conforming 'yes' in a non-confirming 'sometimes' format. Furthermore, saying 'fordi at jeg ved det er billigere'/'because i know it is cheaper', Tine displays that she sometimes drives economically because she *knows* that it is cheaper to do so, thereby pre-empting the municipal suggestion. Hence, simultaneously turning her type-conforming 'yes' into a non-conforming 'sometimes' and audibly stressing 'ved'/'know', Tine reveals that it is of *primary* relevance that she already knows what form of driving is the most economic, *not* how she acts upon this knowledge.

Tine's display of knowledge is extended and justified in lines fifty-two to fifty-seven (it should be noted that line fifty-one is inaudible). In line fifty-two, the turn-initiating 'det' and the short word 'også' show that the turn is a qualification of the former (unfortunately, the English syntax does not follow the Danish here; the English 'it' is placed at the end of the line, but the co-constituting relationship between 'also' and 'it' is of course the same). Primarily, however, lines fifty-two to fifty-seven are tied to the preceding lines simply by being treated as qualification that Tine *knows* it is cheaper to drive in the suggested way.

In greater detail, Tine argues that she has an autonomous *right* to draw on the knowledge that it is cheaper to drive slower. She does so in a piece of category work in which she draws on the membership device 'family'. To establish the knowledge as *her* knowledge, she says that her son has been testing whether it is more economical to drive a hundred and twenty (than, naturally, to drive faster), thereby invoking her own membership of a parent category. Due to the duplicative organisation of the family device (and Sacks's consistency rule), the possessive pronoun 'min'/'my' (line fifty-two) readily invokes a parent category for the speaking part, as well as, by implication, the family device. Thus, Tine ascribes to a collectivity that possesses the knowledge and, consequently, has an autonomous right to it. If, then, Tine acts in accordance with this knowledge, her actions will be premised on a corpus of knowledge held by a collectivity to which she is a member; she will be acting on her own behalf rather than imitating the ways and doings of some other collectivity.

Concluding this part of the analysis, the second pair part can be seen as a non-confirming preferred response to the question concerning the fuel-speed relationship. Yet, since Tine primarily focuses on her autonomous right to the *knowledge* about the relationship, she might be negotiating the request with which she displays alignment. On the one hand, the proffer focused on whether a municipal piece of information could be consequential in relation to actual speed decisions. On the other hand, Tine focuses primarily on her individual right to act in accordance with an already possessed piece of knowledge. Thus, treating her own autonomy as being of primary importance, Tine negotiates the proffer, thereby not only negotiating the imposition of a governmentality discourse (from the council/via the moderator), but also negotiating a discourse of individualism.

In lines sixty and sixty-one, Tine distances herself from the action type alignment of the preceding account. Underlining that 'it' (driving according to the fuel/speed rationale one might hear) is not something she always does, Tine mitigates the preceding. Literally, she merely says, 'men altså ikke noget jeg sådan hver gang overhovedet'/'but well not like something i every time at all'; she does not mention a specific activity. However, the lines readily display a negotiating stance to the preceding. The hesitating turn-initiating inhalation, the disjunctive connective 'men'/'but' and the typical disagreement mitigating 'altså'/'well' display dissociation with the preceding and in doing so links the lines to the preceding. Accordingly, Tine's project comes off as a matter of saying that she does not always drive in accordance with her knowledge and as a matter of weakening the alignment of stance in line forty-seven. Furthermore, this weakening is displayed as *reached for*. In line sixty-one 'hver'/'every' is audibly stressed, and the modal adverb 'overhovedet'/'at all' is remarkably categorical. In combination, the two features highlight that Tine *definitely* does not always drive in a way that (directly) reflects her knowledge of the fuel-speed relationship. Again recalling

Sharrock, the way Tine stresses that she does not always drive according to her knowledge could be heard as revealing that her knowledge entails the right, not the obligation, to drive in this specific way.

In lines sixty-three to eighty-six, Tine accounts for her decreased alignment, emphasising her understanding that economic driving cannot be an obligation. The account comprises three parts, and the first of these is represented in lines sixty-three to sixty-five. Tine treats these first lines as an explanation of why she does not always drive 'economically'. Saying that 'det kommer an på igen hvad man skal'/'it again depends on what you have to do', she implies that 'man' (you/one) has to consider other things when deciding how to drive. Specifically, in stressing 'skal'/'have to', Tine displays that 'man' (you/one) might have other obligations, thereby negatively or indirectly invoking that economic driving is not quite compulsory. Hence, 'driving economically' is exactly *not* approached as being a category-bound obligation.[18]

Tine unfolds this category work as she delivers a concrete and personal example in the second part of the account. Specifically, in lines sixty-eight to seventy-five, Tine draws on an example from her own life in order to illustrate the general point that 'man' (you/one) cannot always drive economically.[19] The form of the example is a typical conditional design, featuring an antecedent from lines sixty-eight to seventy and a consequent from the end of line seventy to line seventy-five. The design incorporates an example of a category-bound obligation that evens out any call to drive economically. In lines sixty-eight to seventy, Tine initiates the account by saying, 'på torsdag hvor jeg skal nå en færge kvart over seks så skal jeg videre til nyborg og være til tiden'/'this thursday where i have to catch a ferry a quarter past six then i have to continue to nyborg and be on time'. Then she takes a brief pause but continues from the end of line seventy to line seventy-one (and also, it should be noted, non-verbally in line seventy-two), delivering the remaining part of the antecedent of the account. It is not easy to make exact

18 Or, more specifically, *if* it is in fact bound to a category, then it is bound to a category on a lower level in a hierarchy of relevance and order of consequences. Jayyusi (1984) points out that a range of interactional problems get resolved by setting up a hierarchy of relevance or consequence. She notes that such hierarchies are of course procedurally constructed (66). Against this, economic driving comes off as being of lower hierarchical relevance; that is, it is readily treated as something that is only relevant in cases where there is not *some*thing else to take into consideration.

19 She treats it as a general phenomenon that is concretised through a personal account by shifting from the broad, unspecific pronoun 'man' (you/one) to the precise, personal pronoun 'jeg'/'i'.

'grammatical' sense of this part, but it seemingly displays a hypothetical consideration of how to use the speed pilot ('den'/'it' comes off as referring to the speed pilot in 'jeg sætter den bare på hundrede'/'i just set it at hundred'). Lastly, represented as 'fordi altså så kommer jeg måske for sent ikke'/'because well then i will maybe be late right', the consequent readily displays a concern about being late. Consequently, the account is readily heard as a qualification of why economic and, by local implication, slower driving cannot always be taken into account. Thus, not only does Tine invoke the obligation 'to be on time' to clarify why economic driving is not always an option, the obligation is accomplished as part of an account of decreased alignment with the moderator's orientation to considerations of a fuel/speed relationship.

However, whereas the categorial modality and the ordinary inference structure of the conditional readily invoke the category-bound obligation 'to be on time', no specific, binding category is *explicitly* invoked. As demonstrated, it is not necessary to consider the category to point out the obligation, but it should, nevertheless, be considered whether Tine does in fact invoke and subscribe to an 'employee' category. Accordingly, I would like to point out three interrelated points.

Firstly, in the local context, everyone knows that Tine is going to Nyborg on a job-related course the coming Thursday. This is due to the fact that the present focus group session was initially planned for that Thursday, but then rescheduled because of this course. Secondly, Tine might be heard as referring to an already known case, specifically through the adverb 'hvor'/'where' and, by implication of this choice, the order of subject and verb in the Danish original.[20] Thirdly, the readily utilised availability of the 'be on time' obligation in itself invokes the relevance of precisely an 'employee' category.

Considering the latter point, recall Jayyusi's (1984) observation on categorial hierarchies or relevancies. Jayyusi points out that it is a part of ordinary interaction to solve different sorts of dilemmas by generating hierarchies (138). Against this, the interaction at hand could be understood as a matter of solving the dilemma of driving economically or being on time by generating a hierarchy where the latter is clearly more important. However, even though hierarchical importance is a matter of in situ practical purpose, it is at the same time culturally afforded and affording (138). Not only is the 'employee' category in general

20 If, oppositely, the future event was mentioned for the first time, Tine could have referred to it with the more explaining 'da skal jeg'/'then I have to'. However, whereas this last reference would not do for a second mention, it is not the case that the 'hvor jeg skal' reference would not at all do for a first reference. It would be more remarkable and more dramatic, but not unlikely.

considered to be of the utmost importance in Western cultures of advanced liberalism, the participants to this focus group setting and the participants to the other settings within the scope of this project generally draw on their memberships of such categories as the ultimate explanation and legitimation for all sorts of mobility practices. Hence, overall the category is treated as the category of importance par excellence. Considering this, Tine apparently invokes this category by invoking the obligation to 'be on time'. Finally, observe how effortlessly Tine in fact draws on this obligation. By clearly *not* treating the time of *departure* as being of any relevance in the account (she does not at all consider if she could make sure she was on time by simply leaving home earlier), she can be said to draw on, and therein sustain, a dominating, individualising 'time fetish' within advanced liberalism.

In line seventy-eight Tine goes back to the broad and unspecific personal pronoun 'man' (you/one), marking the beginning of the third part of the account. Hence, in the remaining lines, Tine delivers a 'conclusion' to her response to the moderator's proffer. Line seventy-eight is not carried out, but rather left; however, in line eighty-one, immediately after the moderator's positive response token in line eighty, Tine starts again. She breaks off towards the end of the line and moves in an expressive way, 'rolling' her hand. She continues in line eighty-three, but breaks off and takes yet another rather extended pause. However, from the second half of line eighty-three and to line eighty-six, she delivers her meticulously worked up argument in a rather direct version, saying that 'man' (you/one) cannot just consider the economic aspect, 'man' (you/one) also has to mind things that 'man' (you/one) has to do. Against this, the hedging restarting and pauses display mitigation tokens, projecting a dispreferred response, or, more radically, against this, *all* of the above display mitigation of a final, dispreferred conclusion. At this point, then, Tine's response seems to be presenting a stance that is only vaguely aligned with the stance displayed in the moderator's question.

In line eighty-seven, the moderator accepts the fate of her proffer, and after a lengthy pause, Tine ends her second pair part in line eighty-nine with a closing, whispered concentrate of the account work she has done. Saying 'tingene skal jo ligesom hænge sammen'/'things after all have to sort of fit together', Tine is readily heard to display that the *right* to drive economically and *obligations* to other aspects of life have to fit together. Hence, in the end, Tine distances herself from the moderator's stance, showing that 'you' do not have an obligation to drive economically, but mainly an autonomously obtained right to do so if it does not interfere with 'your' responsibilities. By orienting to the other participants, seemingly asking whether they agree, the moderator fulfils the projected ending of Tine's second pair part in line ninety.

Concluding, even though the moderator worked hard to project a preferred, not morally threatening, response, the second pair part has notably negotiated the suggested relation between economic driving and a fuel-speed relationship. Drawing on a 'freedom as autonomy' doctrine and an imperative of individual choice, Tine has co-accomplished economic driving in a way that has reinforced her freedom as autonomy, simultaneously isolating her from the obligating and restrictive municipal call for greener transportation. Accordingly, throughout the interaction, the obligating call of the campaign has been pre-empted, Tine's freedom has been maintained, and the regimes of rationality that co-constitute automotive mobility have been negotiated in a way that simultaneously incorporated and dismantled concerns about greener transportation.

Summary

This chapter has revolved around three excerpts from focus group settings conducted after the municipal theme month. In these sessions, the moderator generated topic talk on the municipal transportation theme month, thereby making rationalities and moralities which co-constitute the governing of automobility available for enquiry in the face of a more complex call to account for one's own practices in relation to a strategy aimed at changing those practices. Hence, the ambition was that these sessions would display the continually contested intersection of the practices concerned to conduct the conduct of others and the practices concerned to conduct one's own conduct.

The first analysis in Chapter Seven showed that the transference of rational accountability is a feature of a possible rationalisation of the intersection of the municipal attempt to conduct citizens' transportation conduct and the participants' attempt to conduct their own transportation conduct. The analysis demonstrated how the parties to the focus group setting transfer their rational accountability for the strategy to incumbents of other memberships. It revealed that the participants ascribe to memberships that are not accountable for knowing about the transportation theme month and that they simultaneously resist, and thus invoke, memberships that are accountable for knowing, namely memberships bound by predicates such as 'using public transportation' and 'going to work'.

Both the second and third analyses demonstrated that pre-emption of rational accountability is a feature of a possible rationalisation of the intersection of the municipal attempts to conduct citizens' transportation conduct and the participants' attempts to conduct their own transportation conduct. However, the two analyses differed in that the pre-emption was done in two different ways. The second analysis showed how the participants approach the municipal attempt to conduct citizens' transportation conduct as inconsequential, seeing as the participants already comply with the strategy due to their specific memberships. In more detail, the analysis showed how the parties accomplish an in situ 'pre-fulfilment' of the municipal strategy concerning greener driving practices. The analysis showed how green driving is accomplished as advisable, but that it takes considerable negotiation to accomplish that the municipal incitement to do so is of no actual consequence because of their membership of an 'elderly' category; they already drive in the greenest way possible. The third analysis, however, showed that the participants approached the municipal attempt to conduct citizens' transportation conduct as an activity that made no actual difference for the reason that the participants already hold the relevant knowledge of the suggested conduct. The analysis revealed that the participants accomplish their already possessed knowledge of economic driving as appropriate for all practical purposes. Additionally, it showed how the participants accomplish economic driving as an individual choice; that is, as an individually achieved right, not as an obligation.

I would like to suggest that all three analyses provide useful insights into the rationalities that co-constitute the governing of automobility. In particular, I would like to suggest that the analyses show how available mentalities of government co-constitute the maintenance of automobility. Against this, a couple of points should be recapitulated. Firstly, all three analyses have shown that the participants approach the municipal strategy as a governmentality strategy, as a strategy aimed at governing through the conduct of conduct. Secondly, however, this approach has been part of different rationalisations of the contested intersection between the municipal attempt to conduct citizens' conduct and the participants' conduct of their own conduct. In the first analysis, the transference of accountability meant that the participants withdrew themselves from negotiations at the intersection. Thereby, the rationality of the strategy was de facto limited to a rationality in which proper conduct of transportation conduct in the face of climate change is reduced to a matter which only concerns people who already use public transportation, for instance, when they commute to work or school. In both the second and third analyses, the governmental strategy was fulfilled in and through the pre-emption of it. However, in both analyses, the participants 'took advantage' of the governmental utilisation of their freedom by subtly negotiating the rationality of the governmental strategy. Even though this negotiation might be more readily observable in the last analysis, in which Tine skilfully accomplishes economic driving as a right, not an obligation, even Svend's (and Carin's) 'mere' pre-emption implies a negotiation of the pre-empted rationality. The 'elderly' category not only pre-empts the municipal strategy aimed at green driving; it also accomplishes it. In and through the utilisation of the 'elderly' category as a pre-emption resource, green driving is accomplished as nothing beyond how members of this category membership already drive.

Consider again that these findings are based on the analysis of only a few instances. However, both the transferring rationality and the pre-emption rationality appear to be utilised quite often in the larger corpus of the study (see Lindegaard, 2012). Considering transference, there is, for example, an instance in which a participant transfers accountability, saying that she does not need information on green driving, but that younger and older drivers might. Similarly, a couple of retired men make their age observably relevant in their accounts for their preference of having a car of their own, thereby indirectly transferring accountability for considering shared cars and public transportation to younger people. Thirdly, a participant quite simply refers accountability for using the bus to those who depend on the bus, and, lastly, the case study points to an instance in which two women deliver similar accounts, transferring accountability to people who live in other, less dispersed areas. Considering the pre-emption rationality, the case study touches upon numerous other instances in which participants pre-empt the municipal strategy. I am not aware of any other instances in which it is done as in the analysis above, namely by direct reference to category membership. However, there seems to be quite a few instances in which participants pre-empt the strategy through reference to their own already existing 'dispositions', as such rendering the municipal strategy less relevant. For instance, the case study comprises three excerpts in which two different women invoke their own and other's personal feelings and mentalities as reasons why they and others drive as they do, and it comprises five excerpts in which three different participants all, though very differently, show that they already hold the suggested knowledge about proper conduct of transportation conduct.[21]

21 Considering the question of generalizability, please see the discussion in the end of Chapter Six.

8. Reconsidering rationalities of automobility

By connecting up, on the one hand, the theoretical insight that current forms of governing rely on the projection and maintenance of certain governmental rationalities with, on the other hand, an ethnomethodological approach to discourse analysis, this book has investigated the ongoing negotiation of governmental rationalities of car-dependent life. Particularly, the book has provided analyses of focus group interviews with citizens stemming from sessions before and after the execution of a municipal transportation strategy. The analysis of sessions prior to the municipal strategy showed how the governing of automobility is currently unfolded at the molecular level, namely at the level of accounting for and, hence, conducting one's own transportation conduct. Following the observation that current forms of governing are bound up with a specific utilisation of the freedom of the governed, the analysis demonstrated how biking is rationalised as a free choice and how the obligation to go by car is invoked to account for an already maintained practice. In addition, the analysis of sessions carried out after the strategy showed how the municipal strategy is accomplished as a governmentality strategy, a strategy aimed at governing through the conduct of conduct. However, the analysis showed how this happens through subtle negotiation of the intersection between the municipal attempt to conduct citizens' conduct and citizens' conduct of their own conduct. It showed how the municipal strategy is limited either to a rationality in which proper conduct of transportation conduct in the face of climate change is reduced to a matter of other people already using public transportation or to a rationality in which the proper conduct of transportation conduct is reduced to a matter of maintaining the conduct already autonomously conducted by the citizens. Hence, drawing on both studies of governmentality and ethnomethodology, the book has pointed out how villagers in a small Danish village negotiate a municipal strategy aimed at conducting their transportation conduct in ways that merely sustain the villagers' already maintained car-dependent life forms.

Based in its theoretical argument and empirical analyses, the book contributes to two research domains, namely, firstly, studies of governmentality and, secondly, mobilities research. At first considering studies of governmentality, the book demonstrates how EMCA can be utilised in order to enquire into the continuous accomplishment and negotiation of governmental rationalities at the molecular level. Generally, studies of governmentality point out that it is futile to assume universal categories, such as universal forms of human subjects. For instance,

Miller and Rose (2008) warn us not to presuppose the prevalence of any one kind of subject within current governmentalities, and they call for research on the techniques in play "when individuals judge and evaluate themselves and their lives, sought to master, steer, control, save or improve themselves" (7). Relatedly, studies of governmentality emphasise that it is important to pay attention to *how* governing is unfolded at the *molecular* level. Rose (1999) clarifies this point when he observes that it is not enough to direct attention to what is often considered discursive 'heavy-weights', such as grand official documents, but that also the discursive interactions of *vulgar everyday* life should be considered. "[I]t is, most often, at this vulgar, pragmatic, quotidian and minor level that one can see the languages and techniques being invented that will reshape understandings of the subjects and objects of government, and hence reshape the very presuppositions upon which government rests" (31). Furthermore, Dean (2010) recognises the importance of discourse in his observation that the moral side of government is observable in people's accounts of their conduct. However, whereas Rose and Dean in different ways explicate the same general ambition to pay attention to how rationalities of government are accomplished in discursive interaction at the molecular level, this ambition is not fully fulfilled in current studies of governmentality due to the fact that such studies have not yet connected their broader, historical analyses with analysis of actual discursive interaction at the molecular level. Furthermore, it seems to be the case that when studies of governmentality approach discourse, it is to grasp what *presupposed* concepts make possible and do (cf. Dean, 2010: 79). In contrast, ethnomethodology offers a set of methods uniquely sensitive to the mutually constitutive relationship of power and resistance in the continuous accomplishment of rationalities of government, affording enquiries into the discursive interaction at the molecular level that escape the risk of presupposing the importance of any concepts.

In greater detail, the analysis undertaken in this book has not been conducted with the ambition of investigating, for instance, what the concept of 'freedom' makes possible. I am not even sure that the word 'freedom' ('frihed') is ever used in the data. On the other hand, I have been able to point out the in situ observable relevance of categories such as 'elderly' and 'car owner', of the possession of certain knowledge and of different obligations and rights constituted by and constituting the different categories and ownerships; and I have shown how these 'concepts' have been invoked to accomplish certain things. For instance, I have shown how certain categories can be utilised to pre-empt a municipal strategy and, therein, to negotiate the contested intersection of a municipal strategy aimed at conducting citizens' conduct and the participants' attempt to conduct their own conduct. In this way, without looking for the concept of freedom, the analysis has shown the accomplishment of a rationality which can be interpreted in Rose's terms of freedom as itself a set of powers. Furthermore, the analysis has shown that this set of powers is dispersed and utilised at *all* levels or, in other words, that the relationship between governmental strategies and resistance is particularly contradictory as resistance marks both a boundary and a constitutive moment of government. This could be seen as both a specific and a general contribution to studies of governmentality. Firstly, I propose that this book contributes to Rose's imminent historical study of the powers of freedom in that it shows how freedom is currently utilised at the molecular level. Secondly, I propose that the book has pointed to how studies of governmentality can refine its analysis of discursive interaction at the molecular level in a way that takes seriously the risk of presupposing one's own findings and, thereby, duplicating neoliberal governmentality (cf. Bröckling et al., 2011). Acknowledging the assumption that it is necessary to pay attention to how power is dispersed into and resisted in discourses at the vulgar, everyday level, I do not consider this an insignificant contribution.

Considering, secondly, mobilities research, the book has suggested that in order to properly understand current forms of car-dependent mobility in the face of climate change, it is necessary to go beyond a systems-oriented approach to automobility and focus also on how the governing of car-dependent living is mutually constituted by the ways in which people rationalise their own and others' transportation practices in order to accomplish their conduct of conduct. Particularly, the book contributes to mobilities research in two interrelated ways. Firstly, whereas the book represents neither the only nor the first orientation to rationality in enquiries into the relationship between climate change problematics and automobility, it contributes significantly to this as yet underdeveloped aspect of mobility by taking quite a different approach to rationalities than the one taken in earlier studies. For instance, whereas Freudendal-Pedersen (2009) approaches rationality as a phenomenon that can be mapped out through an investigation of 'structural stories', this book does not approach rationality as something that is already there, but rather as a continuous accomplishment, and, as a consequence, the book does not hypostasise any of its findings, but points out how members continuously accomplish and negotiate rational orders of mobility in their talk about their everyday transportation practices. Secondly, by demonstrating how governmental rationalities of mobility are accomplished in categorial practices, the book might be contributing to the area of mobilities research concerned with the phenomenon of potential mobility (Kaufmann 2002; Kaufmann, Bergman & Dominique, 2004; Kaufmann and Montulet 2008; Sheller 2011; Urry 2007). That is, as the book understands governmental rationalities of mobility as accomplished in discursive interaction (in and through, for instance, people's interpretation and legitimation of various access and skills) it could be understood as being concerned with appropriation and, by implication, the book could be understood as contributing with a useful analytical operationalization in regards to the slowly emerging field of research concerned with potential mobility or, with Kaufmann, with motility.

Concluding, I would like to suggest that the book has indicated the possible shortcomings of strategies in which governors aim to conduct the conduct of the governed through the utilisation of the abilities of the latter to act differently. In the face of climate change, this mode of governing might be somewhat deficient, seeing as the accomplishment of the government of automobility comes off as reinforcing the rationalities of already existing conduct. In other words, it appears to be the case that in and through the accomplishment of the municipal strategy as a governmentality strategy, governing automobility becomes an act of maintaining automobility. Consequently, it might be difficult to get beyond automobility, if it is considered of primary importance that nobody recognises that they themselves might in fact be the ones called upon to conduct their conduct in a way that differs from the ways in which they currently do so.

Bibliography

Austin, J. L. (1962). *How to Do Things With Words.* (J. O. Urmson, Ed.) Oxford: Clarendon Press.

Baker, C. (2004). Membership categorization and interview accounts. In D. Silverman (ed.), Qualitative Research: *Theory, Method and Practice* (pp. 162–176). London, Thousand Oaks and New Delhi: Sage Publications.

Banister, D., Schwanen, T., Anable, J. (eds.). (2012). Theoretical perspectives on climate change mitigation in transport. Special Section. *Journal of Transport Geography*, 24.

Barber, B. R. (2007). *Consumed: How Markets Corrupt Children, Infantilize Adults and Swallow Citizens Whole.* New York: Horten & Company.

Biebricher, T. (2008). Genealogy and Governmentality. *Journal of the Philosophy of History* 2, 363–396.

Böhm, S., Jones, C., Land, C., Paterson, M. (eds.). (2006). *Against Automobility.* Malden, Oxford and Victoria: Blackwell Publishing.

Bogen, D., & Lynch, M. (1990). Social critique and the logic of description: a response to McHoul. *Journal of Pragmatics*, 14, 505–21.

Bröckling, U., Krasmann, S., & Lemke, T. (eds.) (2011). *Governmentality. Current Issues and Future Challenges.* New York and Abingdon, Oxon: Routledge.

Button, G. (1991). Introduction: ethnomethodology and the foundational re-specification of the human sciences. In G. Button (ed.), *Ethnomethodology and the human sciences* (pp. 1–9). Cambridge: Cambridge Universtity Press.

Conley, J. & Tigar McLaren, A. (eds.). (2009). *Car Troubles.* Abingdon, Oxon: Ashgate Publishing Group.

Coulter, J. (1979). *The Social Construction of Mind.* London and Basingstoke: Macmillan Press Ltd.

Coulter, J. (1991). Logic: ethnomethodology and the logic of language. In G. Button (ed.), *Ethnomethodology and the human sciences* (pp. 20–50). Cambridge: Cambridge University Press.

Crosta, N., Davies, A., & Maguire, K. (2006). *The New Rural Paradigm: Policies and Governance.* Paris: OECD Publishing.

Dean, M. M. (2010). *Governmentality: Power and Rule in Modern Society* (2nd Edition ed.). London: Sage publications Ltd.

Dreyfus, H. L., & Rabinow, P. (1983). *Michel Foucault: Beyond Structuralism and Hermeneutics* (2nd Edition ed.). Chicago: The University of Chicago Press.

Dudley, G., Geels, F., & Kemp, R. (2011). *Automobility in Transition? A Socio-Technical Analysis of Sustainable Transport*. New York and London: Routledge.

Eglin, P., & Hester, S. (1999). 'You're all a bunch of feminists!': Categorization and the politics of terror in the Montreal Massacre. *Human Studies*, 22, 253–272.

Fitzgerald, R. (2012). Membership categorization analysis: Wild and promiscuous or simply the joy of Sacks? *Discourse Studies*, 14 (3), 305–311.

Foucault, M. (1972). *The Archaeology of Knowledge & The Discourse on Language*. New York: Pantheon Books.

Foucault, M. (1980). *Power/Knowledge: Selected Interviews and Other Writings 1972–1977*. (C. Gordon, Ed.) London: Harvester.

Foucault, M. (1983). The subject and power. In H. Dreyfus and P. Rabinow, *Michel Foucault: beyond structuralism and hermeneutics*, Chicago: University of Chicago Press, pp. 208–226.

Foucault, M. (2007). *Security, Territory, Population. Lectures at the Collège de France 1977–78*. New York: Palgrave.

Foucault, M. (2007b). *The Poltics of Truth*. (S. Lotringer, Ed.) Los Angeles: Semiotext(e).

Freudendal-Pedersen, M. (2009). *Mobility in Daily Life: Between Freedom and Unfreedom*. Farnham, UK and Burlington, VT: Ashgate.

Freund, P., & Abrams, M. (1976). Ethnomethodology and Marxism: Their Use for Critical Theorizing. *Theory and Society*, 3 (3), 377–393.

Gardner, R. (2012). Enriching CA through MCA? Stokoe's MCA keys. *Discourse Studies*, 14 (3), 313–319.

Garfinkel, H. (1963). A conception of, and experiments with, 'trust' as a condition of stable concerted actions. In O. J. Harvey, *Motivation and Social Interaction* (pp. 187–238). New York: Ronald Press.

Garfinkel, H. (1967). *Studies in Ethnomethodology*. Cambridge: Polity Press.

Garfinkel, H. (2002). *Ethnomethodology's Program: Working Out Durkheim's Aphorism*. (A. W. Rawls, Ed.) Lanham: Rowman & Littlefield Publishers, Inc.

Garfinkel, H., & Sacks, H. (1970). On Formal Structures of Practical Actions. In J. C. McKinney & E. A. Tiryakin (eds.), *Theoretical Sociology* (pp. 337–366). New York: Meredith Corporation.

Goodwin, C. (1980). Restarts, Pauses, and the Achievement of Mutual Gaze at Turn-Beginning. *Sociological Inquiry*, 50, 272–302.

Goodwin, C. (1981). *Conversational Organization: Interaction Between Speakers and Hearers*. New York: Academic.

Grancea, L. (2010). Ethnic solidarity as interactional accomplishment: An analysis of interethnic com-plaints in Romanian and Hungarian focus groups.

Discourse & Society 21(2), 161–188. Have, P. T. (1999). *Doing Conversation Analysis*. London: Sage Publications Ltd.

Have, P. T. (1999). *Doing Conversation Analysis*. London: Sage Publications Ltd.

Have, P. T. (2002). Ontology or methodology? Comments on Speer's 'natural' and 'contrived' data: a sustainable distinction? *Discourse Studies*, 4, 527–530.

Hayano, K. (2011). Claiming epistemic primacy: yo-marked assessments in Japanese. In T. Stivers, L. Mondada & J. Steensig (eds.), *The Morality of Knowledge in Conversation* (pp. 58–81). Cambridge: Cambridge University Press.

Heinemann, T. (2005). Where grammar and interaction meet: The preference for matched polarity in response turns in Danish. In A. Hakulinen, & M. Selting (eds.), *Syntax and Lexis in Conversation: Studies on the Use of Linguistic Resources in Talk-in-Interaction* (pp. 375–402). Philadeliphia: John Benjamins Publishing Company.

Heinemann, T. (2006). 'Will you or can't you?': Displaying entitlement in interrogative requests. *Journal of Pragmatics*, 38, 1081–1104.

Heinemann, Trine (2008). Questions of accountability: yes--no interrogatives that are unanswerable. *Discourse Studies*, 10, 55–71.

Heinemann, T. (2009). Two answers to inapposite inquiries. In J. Sidnell (ed.), *Conversation Analysis: Comparative Perspectives* (pp. 159–186). Cambridge: Cambridge University Press.

Heinemann, T. (2010). The question–response system of Danish. *Journal of Pragmatics*, 42, 2703–2725.

Heinemann, T., Lindstöm, A., & Steensig, J. (2011). Addressing epistemic incongruence in question-answer sequences through the use of epistemic adverbs. In T. Stivers, L. Mondada & J. Steensig (eds.), *The Morality of Knowledge in Conversation* (pp. 107–130). Cambridge: Cambridge University Press.

Heritage, J. (1984a). A change-of-state token and aspects of its sequential placement. In J. M. Atkinson, & J. Heritage (eds.), *Structures of social action: Studies in conversation analysis* (pp. 299–345). Cambridge: Cambridge University Press.

Heritage, J. (1984b). *Garfinkel and ethnomethodology*. Cambridge: Polity Press.

Heritage, J. (1988). Explanations as Accounts: A Conversation Analytic Perspective. In C. Antaki (ed.), *Analysing Everyday Explanation: A Casebook of Methods* (pp. 127–44). London: Sage.

Heritage, J. (2011). Territories of knowledge, territories of experience: empathetic moments in interaction. In T. Stivers, L. Mondada & J. Steensig (eds.), *The Morality of Knowledge in Conversation* (pp. 159–183). Cambridge: Cambridge University Press.

Heritage, J., & Atkinson, J. M. (1984). Introduction. In J. Heritage & J. M. Atkinson (eds.), *Structures of Social Action. Studies in Conversation Analysis* (pp. 1–15). Cambridge: Cambridge University Press.

Heritage, J., & Greatbatch, D. (1991). On the Institutional Character of Institutional Talk: The Case of News Interviews. In D. Boden, & D. H. Zimmerman (eds.), *Talk and Social Structure: Studies in Ethnomethodology and Conversation Analysis*. Berkeley: University of California Press.

Heritage, J., & Watson, D. R. (1979). Formulations as conversational objects. In G. Psathas, *Everyday language: studies in ethnomethodology* (pp. 123–63). New York: Irvington.

Hester, S., & Eglin, P. (eds.) (1997). *Culture in Action: Studies in Memberhip Categorization Analysis*. Lanham & London: Internation Institute for Ethnomethodology and Conversation Analysis & University Press of America, Inc.

Hester, S., & Francis, D. (1994). Doing Data: The Local Organization of a Sociological Interview. *British Journal of Sociology*, 45, 675–95.

Hester, S., & Francis, D. (2001). Is Institutional Talk Institutionalised?'. *TEXT*, 20 (3), 391–413.

Hoeven, M. Van der (2012). *CO2 Emissions from Fuel Combustion. Highlights.* Annual report from the International Energy Agency, IEA. Retrieved June 17, 2014 at http://www.iea.org/publications/freepublications/publication/CO2emissionfromfuelcombustionHIGHLIGHTSMarch2013.pdf.

Housley, W., & Fitzgerald, R. (2002). The reconsidered model of membership categorization analysis. *Qualitative Research*, 2, 59–83.

Housley, W., & Fitzgerald, R. (2007). Categorization, interaction, policy, and debate. *Critical Discourse Studies*, 4 (2), 187–206.

Houtkoop-Steenstra, H. (2000). *Interaction and the Standardized Survey Interview: The Living Questionnaire*. Cambridge: Cambridge University Press.

Jayyusi, L. (1984). *Categorization and the Moral Order*. Boston, London, Melbourne and Henley: Routledge & Kegan Paul.

Jayyusi, L. (1991). Values and moral judgement: communicative praxis. In G. Button, *Ethnomethodology and the human sciences* (pp. 227–251). Cambridge: Cambridge University Press.

Jefferson, G. (1972). Side Sequences. In D. Sudnow (ed.), *Studies in Social Interaction* (pp. 295–331). New York: The Free Press.

Jefferson, G. (1985). An exercise in the transcription and analysis of laughter. In T. A. Dijk (ed.), *Handbook of discourse analysis* (Vol. III, pp. 25–34). London: Academic Press.

Jefferson, G. (1989). Preliminary notes on a possible metric which provides for a "standard maximum" silence of approximately one second in conversation.

In D. Roger, & P. Bull (eds.), *Conversation: an interdisciplinary perspective* (pp. 166–96). Clevedon: Multilingual Matters.

Jefferson, G. (1996). A case of transcriptional stereotyping. *Journal of Pragmatics*, 26, 159–70.

Kaufmann, V. (2002). *Re-thinking mobility*. Burlington: Ashgate.

Kaufmann, V., Bergman, M. M., & Dominique, J. (2004). Motility: Mobility as Capital. *International Journal of Urban and Regional Research*, 28 (4), 745–56.

Kaufmann, V., & Montulet, B. (2008). Between social and spatial mobilities: The issue of social fluidity. In W. Canzler, V. Kaufmann, & S. Kesselring (eds.), *Tracing Mobilities: Towards a Cosmopolitan Perspective* (pp. 37–56). Farnham and Burlington: Ashgate.

Klausen, J. Z. (2012, 01 10). *Transnational Living in Everyday Practices: A Study of Social and Discursive Aspects of Transnational Networking and its Role in Identity Construction*. Retrieved 02 06, 2012 from Aalborg University. VBN Publication: http://vbn.aau.dk/en/publications/transnational-living-in-everyday-practices%28a8b0fe8e-f667-4d3b-ab60-16329449c6f0%29.html.

Laclau, E., & Mouffe, C. (1985). *Hegemony and Socialist Strategy*. London: Verso.

Land of Opportunity, project summary. (2009, 12). Retrieved 07 03, 2012 from Mulighedernes Land: http://www.mulighedernesland.dk/#/41956/.

Latour, B. (2005). *Reassembling the Social: An Introduction to Actor-Network-Theory*. New York: Oxford University Press.

Laurier, E., & Philo, C. (2004). Ethnoarchaeology and undefined investigations. *Environment and Planning A*, 36, 421–436.

Lemke, T. (2002). Foucault, Governmentality, and Critique. *Rethinking Marxism*, 14, 9–64.

Lepper, G. (2000). *Categories in Text and Talk*. London, Thousand Oaks and New Delhi: Sage Publications Ltd.

Lindegaard, L. B. (2010). Searles talehandlingsteori – fra sandhedsbetingelser til tilfredsstillelsesbetingelser. *Res Cogitans* 7 (1), 48–73.

Lindegaard, L. B. (2012) *Automobility at the interface of discourse and governmentality: A study of the accomplishment of rationalities that co-constitute car-dependent mobility in the face of climate change*. Unpublished PhD-dissertation, Aalborg University.

Lindegaard, L. B. (2014) Doing focus group research: Studying rational ordering in focus group interaction. *Discourse Studies*, DOI: 10.1177/1461445614538563.

Lynch, M. (1993). *Scientific practice and ordinary action*. Cambridge: Cambridge University Press.

Lynch, M. (2000). The Ethnomethodological Foundation of Conversation Analysis. *TEXT*, 20 (4), 517–32.

Lynch, M., & Bogen, D. (1994). Harvey Sacks's Primitive Natural Science. *Theory, Culture and Society*, 11 (1), 65–104.

Lynch, M., & Bogen, D. (1996). *The Spectacle of History: Speech, Text and Memory at the Iran-Contra Hearings*. Durham: Duke University Press.

Malpas, J., & Wickham, G. (1995). Governance and failure: on the limits of sociology. *Journal of Sociology*, 31 (3), 37–50.

Mazeland, H., & Have, P. T. (1996). Essential Tensions in (Semi-)open Research Interviews. In I. Maso & F. Wester (eds.), *The Deliberate Dialogue: Qualitative Perspectives on the Interview*. Brussels: VUB University Press.

McHoul, A. (1986). The Getting of Sexuality: Foucault, Garfinkel and the Analysis of Sexual Discourse. *Theory Culture Society*, 3, 65–79.

McHoul, A. (1988). Language and the sociology of mind: a critical introduction to the work of Jeff Coulter. *Journal of Pragmatics*, 12, 229–86.

McHoul, A. (1990). Critique and description: an analysis of Bogen and Lynch. *Journal of Pragmatics*, 14, 523–32.

McHoul, A. (1994). Towards a Critical Ethnomethodology. *Theory Culture Society*, 11, 105–126.

McHoul, A. (1996). *Semiotic Investigations. Towards an Effective Semiotics*. Lincoln & London: University of Nebraska Press.

McIlvenny, P. (2008). A Home Away from Home: Mediating Parentcraft and Domestic Space in a Reality TV Parenting Programme. Home Cultures: The Journal of Architecture, *Design & Domestic Space*, 5 (2), 141–166.

McIlvenny, P. (2009). Communicating a 'Time-out' in Parent-Child Conflict: Embodied Interaction, Domestic Space and Discipline in a Reality TV Parenting Programme. *Journal of Pragmatics*, 41 (10), 2017–2032.

Miller, G. (2004). Building Bridges: The Possibility of Analytic Dialogue Between Ethnography, Conversation Analysis and Foucault. In D. Silverman (ed.), *Qualitative Research: Theory, Method and Practice*. London: Sage.

Miller, P., & Rose, H. (2008). *Governing the Present*. Cambridge (UK), Malden (US): Polity.

Packer, J. (2008). *Mobility without Mayhem: Cars, Safety, and Citizenship*. Durham: Duke University Press.

Paterson, M. (2007). *Automobile Politics: Ecology and Cultural Political Economy*. Cambridge: Cambridge University Press.

Paterson, M., & Stripple, J. (2010). My Space: governing individuals' carbon emissions. *Environment and Planning D: Society and Space*, 28, 341–362.

Peräkylä, A. (2004). Reliability and validity in research based on naturally occurring social interaction. In: Silverman D (ed.) *Qualitative Research: Theory, Method and Practice*. London: SAGE.

Plunkett, R. (2009). Fashioning the feasible: Categorization and social change. *Australian Journal of Communication*, 36 (3), 23–44.

Pollner, M. (1987). *Mundane reason*. Cambridge: Cambridge University Press.

Pomerantz, A. (1980). Telling my side: 'limited access' as a 'fishing' device. *Sociological Inquiry*, 50, 186–198.

Potter, J. (1997). Discourse Analysis as a Way of Analysing Naturally Occurring Talk. In D. Silverman (ed.), *Qualitative Research: Theory, Method and Practice* (pp. 144–160). London: Sage.

Potter, J., & Wetherell, M. (1995). Natural Order: Why Social Psychologists Should Study (A Constructed Version Of) Natural Language, and Why They Have Not Done So. *Journal of Language and Social Psychology*, 14 (1–2), 216–22.

Publikationer, På vej i mulighedernes land. (2008). Retrieved 07 03, 2012 from Mulighedernes Land: http://www.mulighedernesland.dk/#/483956/.

Puchta, C., & Potter, J. (1999, 3). Asking Elaborate Questions: Focus Groups and the Management of Spontaneity. *Journal of Sociolinguistics*, 324–35.

Puchta, C., & Potter, J. (2004). *Focus Group Practice*. London: Sage Publications Ltd.

Raffnsøe, S., Gudmand-Høyer, M., & Thaning, M. (2008). *Foucault*. Frederiksberg: Samfundslitteratur.

Rapley, T. J. (2001). The art(fulness) of open-ended interviewing: some considerations on analysing interviews. *Qualitative Research*, 1 (3), 303–323.

Rapley, T. (2012). Order, order: A 'modest' response to Stokoe. *Discourse Studies*, 14 (3), 321–328.

Rawls, A. W. (2002). Editor's Introduction. In H. Garfinkel, & A. W. Rawls (Ed.), *Ethnomethodology's Program: Working Out Durkheim's Aphorism* (pp. 1–65). Lanham: Rowman & Littlefield Publishers, Inc.

Rawls, A. W. (2005). Garfinkel's Conception of Time. *Time & Society*, 14 No. 2/3, 163–190.

Raymond, J. (2003). Grammar and Social Organization: Yes/No Interrogatives and the Structure of Responding. *American Sociological Review*, 68 (6), 939–967.

Richards, K., & Seedhouse, P. (eds.) (2005). *Applying Conversation Analysis*. Hampshire and New York: Palgrave MacMillan.

Rose, N. (1999). *Powers of Freedom. Reframing Political Thought*. Cambridge: Cambridge University Press.

Rose, N., & Miller, P. (1992). *Political Power beyond the State: Problematics of Government*. The British Journal of Sociology, 43 (2), 173–205.

Sacks, H. (1992a). *Lectures on conversation* (Vol. 1). (G. Jefferson, Ed.) MA, USA: Blackwell Publishing.

Sacks, H. (1992b). *Lectures on conversation* (Vol. 2). (G. Jefferson, Ed.) MA, USA: Blackwell Publisher.

Sacks, H. (1984). Notes on methodology. In: Atkinson, J. M. and Heritage, J. (eds.) *Structures of Social Action. Studies in Conversation Analysis*. Cambridge: Cambridge University Press.

Sacks, H., & Schegloff, E. A. (1979). Two Preferences in the Organization of Reference To Persons In Conversation and their Interaction. In G. Psathas (ed.), *Everyday Language: Studies in Ethnomethodology* (pp. 15–21). New York: Irvington Publishers.

Sacks, H., Schegloff, E. A., & Jefferson, G. (1974). A Simplest Systematics for the Organization of Turn-Taking for Conversation. *Language*, 50 (4), 696–735.

Schegloff, E. A. (1972). Notes on a conversational practice: formulating place. In D. Sudnow (ed.), *Studies in social interaction*. New York: Free Press.

Schegloff, E. A. (1992). Introduction. In H. Sacks, *Lectures on Conversation* (Vol. I, pp. ix–lii). Oxford: Blackwell Publishing.

Schegloff, E. A. (2007a). *Sequence Organization in Interaction*. Cambridge: Cambridge University Press.

Schegloff, E. A. (2007b). A tutorial on membership categorization. *Journal of Pragmatics*, 39, 462–82.

Schwanen, T., Banister, D., & Anable, J. (2011). Scientific research about climate change mitigation in transport: A critical review. *Transportation Research Part A* 45, 45, 993–1006.

Searle, J. R. (1969). *Speech acts*. Cambridge: Cambridge University Press.

Searle, J. R. (1979). *Expression and meaning*. Cambridge: Cambridge University Press.

Seiler, C. (2008). *Republic of Drivers: A Cultural History of Automobility in America*. Chicago: University of Chicago Press.

Sharrock, W. W. (1974). On Owning Knowledge. In R. Turner (ed.), *Ethnomethodology* (pp. 45–53). Harmondsworth: Penguin.

Sheller, M. (2011). *Mobility*. Sociopedia.isa.

Sheller, M., & Urry, J. (2000). The City and the Car. *International Journal of Urban and Regional Research*, 24 (4), 737–757.

Silverman, D. (1998). *Harvey Sacks*. New York: Oxford University Press.

Silverman, D. (2001). *Interpreting Qualitative Data: Methods for Analysing Talk, Text and Interaction* (2nd Edition ed.). London: Sage.

Silverman, D. (2012). Beyond armed camps: A response to Stokoe. *Discourse Studies*, 14 (3), 329–336.

Silverman, D. (2013). *Doing Qualitative Research*. Sage: London.

Sommer, P. (n.d.). Realdania. English Overview. Retrieved 08 15, 2012 from Realdania: http://www.realdania.dk/English.aspx.

Speer, S. A. (2002a). 'Natural' and 'contrived' data: a sustainable distinction? *Discourse Studies*, 4, 511–522.

Speer, S. A. (2002b). Transcending the 'natural'/'contrived' distinction: a rejoinder to ten Have, Lynch and Potter. *Discourse Studies*, 4, 543–548.

Stivers, T. (2008). Stance, alignment, and affiliation during storytelling: when nodding is a token of affiliation. *Research on Language and Social Interaction*, 41 (1), 31–57.

Stivers, T., Mondada, L., & Steensig, J. (eds.) (2011). *The Morality of Knowledge in Conversation*. Cambridge: Cambridge University Press.

Stokoe, E. (2012). Moving forward with membership categorization analysis: Methods for systematic analysis. *Discourse Studies*, 14 (3), 277–303.

Summerfield, T., & McHoul, A. (2005). Family as a commonsensical device and its place in law. *International Journal for the Semiotics of Law*, 18, 243–261.

Svennevig, J. (2000). *Getting Acquainted in Conversation*. Philadelphia, PA: John Benjamins Publishing Company.

Tate, S. A. (2007). Foucault, Bakhtin, Ethnomethodology: Accounting for Hybridity in Talk-in-interaction. *Forum for Qualitative Social Research*, 8 (2).

Turner, R. (ed.). (1974). *Ethnomethodology*. Harmondsworth: Penguin.

Urry, J. (2004). The 'System' of Automobility. *Theory Culture Society*, 21, 25–39.

Urry, J. (2007). *Mobilities*. Cambridge: Polity Press.

Urry, J. (2008). Climate change, travel and complex futures. *British Journal of Sociology*, 59, 261–279.

Urry, J. (2011). *Climate Change and Society*. London: Polity.

Watson, D. R. (1978). Categorization, Authorization and Blame-Negotiation in Conversation. *Sociology* 12(1), 105–113.

Watson, D. R. (1994). Harvey Sacks's Sociology of Mind in Action. *Theory, Culture and Society*, 11, 169–86.

Watson, D. R., & Weinberg, T. S. (1982). Interviews and the Interactional Construction of Accounts of Homosexual Identity. *Social Analysis*, 11, 56–78.

Watson, R. (1997). Some General Reflections on 'Categorization' and 'Sequence' in the Analysis of Conversation. In S. Hester, & P. Eglin (eds.), *Culture in Action: Studies in Membership Categorization Analysis* (pp. 49–76). Lanham and London: International Institute for Ethnomethodology and Conversation Analysis and University Press of America, Inc.

Watson, R. (2001). The Character of "Institutional Talk": A Reply to Hester and Francis. *TEXT*, 20 (3), 377–91.

Whitehead, K. A. (2012). Moving forward by doing analysis. *Discourse Studies*, 14 (3), 337–343.

Widdicombe, S., & Wooffitt, R. (1995). *The Language of Youth Subcultures*. Hemel Hempstead: Harvester Wheatsheaf.

Wilkinson, S., & Kitzinger, C. (2008). Using conversation analysis in feminist and critical research. *Social and Personality Psychology Compass*, 2 (2), 555–573.

Wittgenstein, L. (1998 [1953]). *Philosophical Investigations*. (G. E. Anscombe, Trans.) London: Blackwell.

Wowk, M. T., & Carlin, A. P. (2004). Depicting a Liminal Position in Ethnomethodology, Conversation Analysis and Membership Categorization Analysis: The Work of Rod Watson. *Human Studies*, 27, 69–89.

Political and Social Change

Edited by Martin Bak Jørgensen and Óscar García Agustín

Volume 1 Martin Bak Jørgensen / Óscar García Agustín (eds.): Politics of Dissent. 2015.

Volume 2 Laura Bang Lindegaard: Congestion. Rationalising Automobility in the Face of Climate Change. 2015.

www.peterlang.com

www.ingramcontent.com/pod-product-compliance
Ingram Content Group UK Ltd.
Pitfield, Milton Keynes, MK11 3LW, UK
UKHW041826031026
15587UKWH00002B/583

9 783631 666630